T0184357

Routledge Revivals

Frege, Dedekind and Peano on the Foundations of Arithmetic

First published in 1982, this reissue contains a critical exposition of the views of Frege, Dedekind and Peano on the foundations of arithmetic. The last quarter of the 19th century witnessed a remarkable growth of interest in the foundations of arithmetic. This work analyses both the reasons for this growth of interest within both mathematics and philosophy and the ways in which this study of the foundations of arithmetic led to new insights in philosophy and striking advances in logic. This historical-critical study provides an excellent introduction to the problems of the philosophy of mathematics - problems which have wide implications for philosophy as a whole. This reissue will appeal to students of both mathematics and philosophy who wish to improve their knowledge of logic.

Frege, Dedekind and Peano on the Foundations of Arithmetic

D. A. Gillies

Routledge
Taylor & Francis Group

First published in 1982
by Van Gorcum

This edition first published in 2011 by Routledge
4 Park Square, Milton Park, Abingdon, Oxon OX14 4RN
605 Third Avenue, New York, NY 10017

Routledge is an imprint of the Taylor & Francis Group, an informa business

© 1982 Van Gorcum & Comp. B.V., P.O. Box 43, 9400 A A Assen, The
Netherlands

Publisher's Note
The publisher has gone to great lengths to ensure the quality of this reprint but
points out that some imperfections in the original copies may be apparent.

Disclaimer
The publisher has made every effort to trace copyright holders and welcomes
correspondence from those they have been unable to contact.

A Library of Congress record exists under ISBN: 9023218884

ISBN 13: 978-0-415-66709-8 (hbk)
ISBN 13: 978-0-203-81628-8 (ebk)
ISBN 13: 978-0-415-66874-3 (pbk)

D. A. Gillies

Chelsea College,
University of London

Frege, Dedekind, and Peano
on the Foundations
of Arithmetic

1982
Van Gorcum, Assen, The Netherlands

© 1982 Van Gorcum & Comp. B.V., P.O. Box 43, 9400 AA Assen, The Netherlands

The publication of this book was made possible through a grant from the Methodology and Science Foundation.

CIP-gegevens

Gillies, D. A. — Frege, Dedekind, and Peano on the foundations of arithmetic / D. A. Gillies. — Assen: Van Gorcum. — Ill.

Met index, lit. opg.
SISO 133   UDC 161/162
Trefw.: wiskunde/logica.
ISBN 90 232 1888 4

Printed in The Netherlands by Van Gorcum, Assen

"... I made the fixed resolve to keep meditating on the question till I should find a purely arithmetic and perfectly rigorous foundation for the principles of infinitesimal analysis."
Dedekind, 1872.

"Your discovery of the contradiction caused me the greatest surprise and, I would almost say consternation, since it has shaken the basis on which I intended to build arithmetic. ... It is all the more serious since, ..., not only the foundations of my arithmetic, but also the sole possible foundations of arithmetic, seem to vanish."
From Frege's letter to Russell, 1902.

Preface

This work contains a critical exposition of the views of Frege, Dedekind, and Peano on the foundations of arithmetic. These views have to some degree been rendered obsolete by discoveries made in the twentieth century — particularly Russell's paradox, Gödel's incompleteness theorems, and Skolem's non-standard models for arithmetic. Despite this, however, the writings of Frege, Dedekind, and Peano still, in my opinion, repay careful study. They are full of most interesting observations, insights and arguments, well worthy of consideration by anyone who today attempts the far from easy task of giving an adequate philosophical theory of arithmetic.

There are, so I believe, advantages in presenting the views of Frege, Dedekind and Peano together, rather than treating Frege in isolation, as is sometimes done. Even if our aim is simply that of understanding Frege, his views will become clearer when they are compared and contrasted with those of some of his distinguished contemporaries. Dedekind, like Frege, was a logicist — that is, he believed that arithmetic could be reduced to logic. However, Dedekind developed this thesis in a different way from Frege. Dedekind regarded the notion of 'set' or 'class' (or, in his own terminology, 'system') as a basic notion of logic. He is thus one of the ancestors of axiomatic set theory, as I try to show by tracing, in detail, Dedekind's influence on Zermelo. Frege, on the other hand, denied that 'set' was a logical notion, and based his logic on the notion of 'concept'. Frege is thus the ancestor of higher-order logic. In contrast to both Dedekind and Frege, Peano denied that arithmetic could be reduced to logic. He is really the forerunner of Hilbert's later formalist philosophy of mathematics.

It is well-known that Frege (and also Peano, independently, though to a lesser degree) made a great advance in logic. I shall argue that this advance arose from investigations into the foundations of arithmetic. The key stimulus was the programme of presenting arithmetic as an axiomatic-deductive system with the underlying logic made fully explicit. It turned out that the underlying logic of such a system for arithmetic was richer than that contained in any previous formal logic.

One attractive feature of the foundations of artihmetic, as an area of study, is that most of the fundamental problems of the philosophy of mathematics appear in this field, but the technicalities involved are less than elsewhere. Since everyone learns arithmetic at school, the only thing unfamiliar to the non-mathematician will probably be the principle of mathematical, or complete, induction, and I have devoted an appendix to explaining this. I hope therefore that any philosophy student, who has done the usual basic course in logic, will be able to read this work without too much difficulty, and thereby gain some knowledge of the problems of the philosophy of mathematics — problems which have, of course, wide implications for philosophy as a whole.

Most of the material which follows was presented in lectures and seminars in the Department of History and Philosophy of Science in Chelsea College, University of London; and I greatly benefited from the penetrating comments and criticisms I received on those occasions. I would particularly like to thank my colleagues Dr. M. Machover and Dr. M. L. G. Redhead who were kind enough to read through the whole thing, and suggested many improvements.

D. A. GILLIES

Chelsea College,
University of London,
Manresa Road, S. W. 3.

Contents

Introduction

The aim of this work is to examine the views of Frege, Dedekind, and Peano on the foundations of arithmetic. By way of introduction, however, it seems wothwhile asking: 'why did these authors get interested in this subject? why did they feel that it would be desirable to provide a firm foundation for arithmetic?' After all, the arithmetic of the natural numbers $\{0, 1, 2, ..., n, ...\}$ had been widely employed by mathematicians in Western Europe since 1500. Why then was it only in the last quarter of the 19th century that serious attempts were made to examine the foundations of the theory of numbers?

I shall not try here to give a complete answer to these questions, but will content myself with mentioning one factor which almost certainly influenced Frege and Dedekind, and probably influenced Peano as well. This factor was the so-called arithmetization of analysis, or, as it might more accurately be described, the definition of real numbers in terms of rational numbers.

The arithmetization of analysis can be dated around 1870. Definitions of real numbers in terms of rational numbers were published by Méray in 1869, and by Cantor, Dedekind and Heine in 1872. Weierstrass had earlier (in the 1860's) expounded a theory of real numbers in his lectures at Berlin.

I shall not here attempt a thorough treatment of the arithmetization of analysis, which is a topic in itself.[1] Instead I shall try to give an informal exposition of what was involved, in order to show how this in turn almost certainly stimulated interest in the foundations of arithmetic. For the purposes of this exposition, I shall use Dedekind's definition of real number as set out in his (1872) Continuity and Irrational Numbers. Dedekind's definition has subsequently proved the most popular of those proposed around 1870,

---

[1] For a good short account see Kline (1972) Mathematical Thought from Ancient to Modern Times. Ch. 41. § 3. pp. 982-987.

and Dedekind's treatment has a particular interest for us, since we are going on later to examine Dedekind's views about arithmetic.

To understand the arithmetization of analysis, we must begin by examining the various kinds of number and how they are related to each other. Simplest of all are the *natural numbers* 0, 1, 2, 3, ..., n, ... We can write:

N = the set of natural numbers = $\{0, 1, 2, 3, ..., n, ...\}$

Here I take the natural numbers as beginning with 0, as Frege does. One can, alternatively, take the natural numbers as beginning with 1, and write:

N = $\{1, 2, 3, ..., n, ...\}$

This is the convention adopted by Dedekind and Peano. It really makes virtually no difference which definition of N is used, and I will adopt whichever is appropriate in a given context. The main body of this work is concerned with natural numbers, and their arithmetic. So, when 'number' is used without further qualification, it can be taken as meaning 'natural number'.

We next introduce Z = the set of integers, where

Z = $\{..., -n, ..., -3, -2, -1, 0, +1, +2, +3, ..., +n, ...\}$.

We can then consider R = the set of rational numbers or fractions. A rational number is one of the form p/q where p, q are integers, and $q \neq 0$.

This account of rational numbers is more or less a definition of rational numbers in terms of integers. We can make this definition a little more precise as follows. Suppose we are given the set Z of integers, we then define a rational number r as an ordered pair $\langle p, q \rangle$ where p, q are integers and $q \neq 0$. To complete this definition, we have first to say when two rational numbers are equal, which we do as follows:

$\langle p, q \rangle = \langle p', q' \rangle$ if and only if $pq' = p'q$

We have then to define the operations of $+$ and $\times$, which can be done thus:

$\langle p, q \rangle + \langle p', q' \rangle = \langle pq' + p'q, qq' \rangle$

$\langle p, q \rangle \times \langle p', q' \rangle = \langle pp', qq' \rangle$

These definitions are obviously obtained by thinking of $\langle p, q \rangle$ as p/q, and applying the usual rules for fractions. It can be verified that $=$, $\times$, $+$ as thus defnied have all the requisite properties.

Just as rational numbers can be defined in terms of integers, so integers can be defined in terms of natural numbers. Indeed we can define an integer p as an ordered pair $[m, n]$ of natural numbers, where we can think of $[m, n]$ as $m - n$. This way of thinking of $[m, n]$

can guide us in defining $=$, $+$, $\times$ for integers, just as thinking of $\langle p, q \rangle$ as p/q guided us in defining $=$, $+$, $\times$ for rational numbers. Since rational numbers can be defined in terms of integers, and integers in terms of natural numbers, we may by combining the definitions, obtain a definition of rational numbers in terms of natural numbers.

Let us now turn to the consideration of real numbers. Perhaps the easiest way into this subject is to begin by observing that rational numbers can be given a geometrical representation. Let us take an arbitrary line 1, and an arbitrary point 0 on it. Take

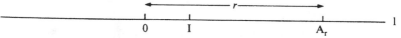

another arbitrary point I to the right of 0. We regard 0I as the unit of distance. Then with any rational number r, we correlate a point A_r on 1 such that the distance $0A_r = |r|$, and A_r is to the right of 0 if $r > 0$, and to the left of 0 if $r < 0$. So to every rational number there corresponds a point A_r on 1. However the converse is not true. There are points of 1 to which no rational number corresponds. This remarkable fact was discovered by the Pythagorean school of ancient Greece. We will next demonstrate its truth with arguments which are probably quite similar to those originally used by the Pythagoreans.

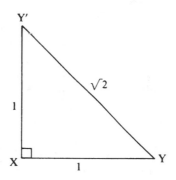

It is an easy matter to construct a triangle XYY′ where Y′X̂Y is a right angle, and $XY = XY' = 0I$ (our unit of length). Now by Pythagoras' theorem $YY'^2 = XY^2 + XY'^2 = 2$. $\therefore YY' = \sqrt{2}$. It is easy to mark off to the right of 0 on 1 a point B such that $OB = YY' = \sqrt{2}$. If we can show that $\sqrt{2}$ is not a rational number, we have obtained a point B to which no rational number corresponds in the given representation.

Suppose then that $\sqrt{2}$ is a rational number i.e. that $\sqrt{2} = p/q$ where p, q are integers and $q \neq 0$. We shall show that this assumption leads to a contradiction, and hence that $\sqrt{2}$ is not rational. For if $\sqrt{2} = p/q$, where $q \neq 0$, each of p, q must be either odd or even, but not both. We can actually suppose that at least one of p, q is odd, for, suppose they are both even, then $p = 2p'$, $q = 2q'$, and $p/q = 2p'/2q' = p'/q'$. If p', q' are again both even, we can cancel again and continue doing so until one of the two becomes odd. But now, since $\sqrt{2} = p/q$, we have

$$p = \sqrt{2}q$$
$$\therefore p^2 = 2q^2$$

$\therefore p^2$ is even. $\therefore p$ is even (since the square of an odd number is odd). $\therefore p = 2p'$

$$\therefore 4p'^2 = 2q^2$$
$$\therefore q^2 = 2p'^2$$

$\therefore q^2$ is even. $\therefore q$ is even. $\therefore$ Both p, q are even. But we have assumed that at least one of p, q is odd. This is a contradiction, and the required result follows.

In view of this result, let us introduce the set **R** of real numbers geometrically. Given any point A on the line 1 (as introduced above), we associate with it a real number α where $|\alpha| = $ the length of 0A, and $\alpha > 0$ if A is to the right of 0 and $\alpha < 0$ if A is to the left of 0. Some real numbers as thus defined are rational numbers, but some (e.g. $\sqrt{2}$) are not. Note that here we are defining real numbers in terms of a geometrical notion viz. length. This is the geometrical approach to the real numbers. It was adopted more or less explicitly by Euclid, and, more or less implicitly, by Western European mathematicians during the period c.1500 - c.1850 when algebra and calculus were being developed. Dedekind begins his (1872) Continuity and Irrational Numbers by an explicit rejection of this geometrical approach to real numbers.

Dedekind recounts that, in the autumn of 1858, as professor in the Polytechnic School in Zürich, he had to lecture on the elements of the differential calculus. He goes on, (1872) Continuity and Irrational Numbers pp. 1-2:

"In discussing the notion of the approach of a variable magnitude to a fixed limiting value, and especially in proving the theorem that every magnitude which grows continually, but not beyond all limits, must certainly approach a limiting value, I had recourse to geometric evidences. Even now such resort to geometric intuition in a first presentation of the differential calculus, I regard as exceedingly useful, from the didactic standpoint, and indeed indispensable, if one does not wish to lose too much time. But that this form of introduction into the differential calculus can make no claim to being scientific, no one will deny. For myself this feeling of dissatisfaction was so overpowering that I made the fixed resolve to keep meditating on the question till I should find a purely arithmetic and perfectly rigorous foundation for the principles of infinitesimal analysis."

The point to notice here is that Dedekind holds that a "recourse to geometric evidences ... can make no claim to being scientific", and that this is something which "no one will deny". In some ways this attitude is strange, since geometry was for many hundreds of years regarded as the most perfect of the sciences. It is certainly interesting to ask why Dedekind had such a mistrust of geometrical considerations. We will not, however, pursue this question here, but rather examine how Dedekind sets up his "purely arithmetic and perfectly rigorous foundation."

Dedekind's problem is to define the set of real numbers (**R**) in terms of the set of rational numbers (R) without using any geometrical notion such as length or magnitude. To do this, he makes use of the key concept of a cut in R. This is defined as follows, Dedekind (1872) Continuity and Irrational Numbers IV pp. 12-13:

"If now any separation of the system R into two classes A_1, A_2, is given which possesses only *this* characteristic property that every number a_1 in A_1 is less than every number a_2 in A_2, then for brevity we shall call such a separation a *cut* [Schnitt] and designate it by (A_1, A_2)."

Every rational number a produces two cuts viz.

$$A_1 = \hat{r}(r \le a) \qquad\qquad A_1 = \hat{r}(r < a)$$
$$A_2 = \hat{r}(r > a) \qquad\qquad A_2 = \hat{r}(r \ge a)[1]$$

Dedekind regards these two cuts as essentially equivalent.

It is clear that the idea of the cut is suggested by geometrical considerations. We can think of a point A on the line 1, cutting 1 into two pieces — one consisting of points to the left of A, and the other of points to the right of A. A itself can be added to either piece. But while the idea of the cut is *suggested* by geometry, the exact definition of cut does not involve anything geometrical. The only notions involved are those of the rational numbers and their arithmetic, and of sets of rational numbers. Thus Dedekind, while using geometry heuristically, has eliminated anything geometrical in his precise treatment.

By the Pythagorean proof given earlier, $\sqrt{2}$ is not a rational number. So the cut

---

[1] The notation which we shall use here and subsequently is fairly standard. It is described in Appendix I, On Notation.

$$A_1 = \hat{r}(\ (r < 0) \ v \ (\ (r \geq 0) \wedge (r^2 \leq 2) \ ) \ )$$
$$A_2 = \hat{r}(\ (r > 0) \wedge (r^2 > 2) \ )$$

is not produced by a rational number. Indeed Dedekind shows that there exist infinitely many cuts not produced by rational numbers. He goes on, (1872) Continuity and Irrational Numbers IV p.15:

> "In this property that not all cuts are produced by rational numbers consists the incompleteness or discontinuity of the domain R of all rational numbers.
>
> Whenever, then, we have to do with a cut (A_1, A_2) produced by no rational number, we create a new, an *irrational* number α, which we regard as completely defined by this cut (A_1, A_2); we shall say that the number α corresponds to this cut, or that it produces this cut. From now on, therefore, to every definite cut there corresponds a definite rational or irrational number, and we regard two numbers as *different* or *unequal* always and only when they correspond to essentially different cuts."

One point to notice here is that Dedekind does not give an explicit set-theoretic definition of the irrational number α in terms of either or both of the sets (A_1, A_2), but rather says that we should "create" the number from the cut. Russell was later to criticize this procedure, and suggested instead that an irrational number be identified with a set of rational numbers. For example α could be identified with the set A_1 of the cut (A_1, A_2).

Dedekind used this idea of "creating" numbers again in connection with the natural numbers. We shall therefore consider his views on this question, and Russell's criticisms in more detail later on (see below Ch. 9 pp. 60-61).

Dedekind next defines a *real* number as a rational or irrational number and proceeds to define relations of order $\alpha < \beta$ and equality $\alpha = \beta$ between real numbers in a fairly obvious way. He is then in a position to state and prove his key theorem on continuity (Section V Theorem IV). This result is formulated as follows, (1872) Continuity and Irrational Numbers p. 20:

> "... the domain **R** possesses also *continuity*; i.e. the following theorem is true:
>
> IV. If the system **R** of all real numbers breaks up into two classes A_1, A_2 such that every number α_1 of the class A_1 is less than every number α_2 of the class A_2 then there exists one and only one number α by which this separation is produced."

This is a key result since it shows that Dedekind's cut construction applied to real numbers does not produce anything new. In this sense the real numbers are complete.

It only remains for Dedekind to define the various elementary operations on real numbers, and to lay the foundations of infinitesimal analysis (that is of the differential and integral calculus). This he does in the last two sections of his monograph (VI and VII).

The only operation which Dedekind treats in detail is addition. Let α and β be real numbers defined by cuts (A_1, A_2) and (B_1, B_2). Then $\gamma = \alpha + \beta$ is defined by a cut (C_1, C_2), where $c_1 \, \varepsilon \, C_1$ if and only if there exist $a_1 \, \varepsilon \, A_1$ and $b_1 \, \varepsilon \, B_1$ such that $a_1 + b_1 \geq c_1$. We have to verify that this definition does define a cut, that it agrees with addition for rational numbers in the case where α and β are rational, and that it gives the standard properties of addition. These are straightforward matters.

Dedekind goes on to say, (1872) Continuity and Irrational Numbers p. 22:

> "Just as addition is defined, so can the other operations of the so-called elementary arithmetic be defined, viz., the formation of differences, products, quotients, powers, roots, logarithms, and in this way we arrive at real proofs of theorems (as, e.g., $\sqrt{2} \, \sqrt{3} = \sqrt{6}$), which to the best of my knowledge have never been established before."

Finally in section VII, Dedekind proves the theorem which he mentioned in the preface viz. (op. cit. pp. 24-25):

> "If a magnitude x grows continually but not beyond all limits it approaches a limiting value."

This, and some other results which he established in Section VII, together provide a basis for analysis (i.e. differential and integral calculus) which is independent of any geometrical considerations. Such then is a brief summary of Dedekind's (1872) monograph. Let us next examine more closely what Dedekind really had achieved in this monograph, and what new problems his results raised.

Dedekind certainly succeeds in defining real numbers in terms of rational numbers, and hence in terms of natural numbers (since rational numbers are easily defined in terms of natural numbers). Moreover he shows how analysis (differential and integral calculus) can be developed from his definition. It thus appears that he has reduced the whole of analysis to a consideration of natural numbers and their arithmetic. This is the meaning of the phrase "arithmetization of analysis". Dedekind himself expresses this point of view in the Preface to his (1888) work: "*Was sind und was sollen die Zahlen?*"[1]

---

[1] Literally: what are and what ought to be the numbers? Since there is no standard English translation of the title of this work of Dedekind's I shall refer to it by its original title. In the same way I shall speak of Frege's (1879) *Begriffsschrift* (literally: Concept writing), and (1893) *Grundgesetze der Arithmetik* (literally: Fundamental Laws of Arithmetic). Some authors do refer to these works by a title translated into English, and, when quoting their views, I have taken the liberty of restoring the original German title in order to avoid confusion.

He writes (op. cit. p. 35):

"... every theorem of algebra and higher analysis, no matter how remote, can be expressed as a theorem about natural numbers, — a declaration I have heard repeatedly from the lips of Dirichlet."

With the advantage of hindsight, we can question this conclusion. Dedekind's definition of real number involves not only the concept of rational number (and hence natural number), but also the concept of *infinite set*. In a Dedekind cut (A_1, A_2), both A_1 and A_2 are infinite sets. Thus analysis is not reduced simply to the theory of natural numbers, but to the theory of natural numbers together with the theory of infinite sets.

It is easy to see why this was overlooked at the time. The notion of set or class must have appeared in the 1870's as straightforward and unproblematic. Indeed Dedekind, as we shall see, regards 'set' (or in his terminology 'system') as a basic notion of logic. Only after Cantor had developed the theory of infinite sets, and the paradoxes of set theory had emerged, did it become clear that set theory is highly problematic and anything but straightforward. From a modern point of view, then, some question-marks hang over the arithmetization of analysis. Dedekind certainly succeeded in eliminating geometrical considerations from the foundations of analysis, but only at the expense of introducing part of the theory of infinite sets. It could be argued, moreover, that the notion of infinite set is as doubtful as, if not more doubtful than, the geometrical notions which it replaced.

This, however, is a later point of view. In the 1870's it certainly seemed that analysis had been successfully reduced to the theory of natural numbers. The obvious next step in constructing a firm foundation for mathematics would be to provide a satisfactory foundation for the theory of natural numbers. Thus there is a natural transition from the arithmetization of analysis in the 1870's to interest in the foundations of arithmetic in the 1880's.

The link here is obvious as far as Dedekind is concerned. His first work on the foundations of mathematics, his (1872) Continuity and Irrational Numbers, is on the arithmetization of analysis, while his second work in the foundations of mathematics, his (1888) *Was sind und was sollen die Zahlen?* deals with the foundations of the theory of natural numbers.

In his (1884) Foundations of Arithmetic, § 1, 1e — § 2, 2e, Frege writes:

"The concepts of function, of continuity, of limit and of infinity have been shown to stand in need of sharper definition. Negative and irrational numbers, which had long since been admitted into science, have had to submit to a closer scrutiny of

8

their credentials... Proceeding along these lines, we are bound eventually to come to the concept of Number and to the simplest propositions holding of positive whole numbers, which form the foundation of the whole of arithmetic."

Here Frege puts the situation into greater historical perspective. The arithmetization of analysis was actually the culmination of a movement, which, beginning with Cauchy early in the 19th century, strove to introduce greater rigour into the development of mathematical analysis. The earlier stages of this "revolution in rigour" did indeed involve a closer consideration of the concepts of function, continuity, and limit; and the whole movement eventually led, via the arithmetization of analysis (or, as Frege puts it, a closer scrutiny of the credentials of the irrational numbers), to an examination of the concept of natural number.

When Dedekind and Frege began their investigations of the foundations of arithmetic, it must have seemed to them that they were completing the last stage in a final and definitive rigorization of mathematical analysis. But the results of their investigations were quite different from what they expected. Instead of providing a firm and lasting foundation for analysis, their works provided an important stimulus for the discovery of basic paradoxes in logic itself. Thus they inadvertently helped to throw the foundations of mathematics into a state of crisis from which they have not really recovered — even today. So the period with which this work deals passes from the certainty with which Dedekind speaks, (1872) Continuity and Irrational Numbers pp. 1-2 of "a purely arithmetic and perfectly rigorous foundation for the principles of infinitesimal analysis" to the uncertainty and confusion displayed in the following extract from Frege's letter to Russell of 1902 pp. 127-8: "Your discovery of the contradiction caused me the greatest surprise and, I would almost say, consternation, since it has shaken the basis on which I intended to build arithmetic. ... I must reflect further on the matter. It is all the more serious since, with the loss of my Rule V, not only the foundations of my arithmetic, but also the sole possible foundations of arithmetic, seem to vanish."

We must beware, however, of attributing an interest in the foundations of arithmetic exclusively to developments in mathematics itself, important though these undoubtedly were. Frege states explicitly that an interest in certain philosophical questions was partly responsible for his undertaking an investigation of the foundations of arithmetic. He writes, (1884) Foundations of Arithmetic § 3 p. 3e:

9

"Philosophical motives too have prompted me to enquiries of this kind. The answers to the questions raised about the nature of arithmetical truths — are they a priori or a posteriori? synthetic or analytic? — must lie in this same direction."

Indeed Frege introduces his own philosophical views about number by giving extended criticisms of the views of other philosophers. The most important of these philosophers are Kant and Mill. We will therefore begin in Ch. 1 by expounding Kant's theory of mathematics, and then give in Ch. 2 Frege's criticisms of Kant. Ch. 3 will give an account of Mill's theory of mathematics, and Ch. 4 will describe Frege's criticisms of Mill. We will then be in a position to begin in Ch. 5 a more detailed account of Frege's own views. Although our concern in this work is with the foundations of arithmetic, we will include in our discussion of Kant and Mill an account of their views on geometry. This will, I believe, help to clarify their over-all positions regarding mathematics.

Chapter 1.
Kant's Theory of Mathematics

Kant's theory of mathematics depends on a pair of distinctions, namely:

(i) between *a priori* and *a posteriori* knowledge, and

(ii) between *analytic* and *synthetic* judgements.

The first distinction is really a traditional one. Kant explains it in (1781) Critique of Pure Reason A2/B3, p. 43 as follows:

> "In what follows, therefore, we shall understand by *a priori* knowledge, not knowledge independent of this or that experience, but knowledge absolutely independent of all experience."

A posteriori knowledge, however, does depend on experience.

The second distinction is really due to Kant himself — though there are traces of it in earlier authors. In (1781) Critique of Pure Reason, A6/B10, p. 48, the distinction is made as follows:

> "Either the predicate B belongs to the subject A, as something which is (covertly) contained in this concept A; or B lies outside the concept A, although it does indeed stand in connection with it. In the one case I entitle the judgment analytic, in the other synthetic."

Kant gives the following example to illustrate the distinction:

Analytic judgement: All Bodies are Extended.

Synthetic judgement: All Bodies are Heavy.

Kant believes that the concept of body contains the concept of extension, so that to talk of an unextended body would involve a contradiction. Thus "all bodies are extended" is analytic. On the other hand, although all bodies are indeed heavy, it is no way contradictory to conceive of a weightless body. Thus "all bodies are heavy" is synthetic.

This example is somewhat doubtful since Boscovich had proposed in 1759 that matter might consist of point atoms with no spatial extension, but acting on each other at a distance. Such point atoms appear to be conceivable, and, were they to exist, they would presumably be bodies of a kind. Thus the judgement "all bodies are extended" cannot, after all, be analytic. A favourite modern example "all bachelors are unmarried" seems better suited to illustrate Kant's notion of analytic judgement.

Frege makes an interesting (and subsequently often repeated) criticism of Kant's distinction between analytic and synthetic. He writes (1884) Foundations of Arithmetic, § 88, p. 100e:

"On the basis of his definition, the division of judgements into analytic and synthetic is not exhaustive. What he is thinking of is the universal affirmative judgement; there, we can speak of a subject concept and ask — as his definition requires — whether the predicate concept is contained in it or not. But how can we do this, if the subject is an individual object? Or if the judgement is an existential one? In these cases there can simply be no question of a subject concept in KANT's sense. He seems to think of concepts as defined by giving a simple list of characteristics in no special order; but of all ways of forming concepts, that is one of the least fruitful."

The case where the subject is an individual object e.g. Socrates is mortal does not really pose serious difficulties, since Kant would presumably say that such judgements are synthetic. However, the case of existential judgements is more problematic.

The trouble is that Kant here, as everywhere else in the Critique of Pure Reason, assumes the correctness of Aristotelian logic. In particular he assumes that all propositions are of the subject-predicate form: S is P. It thus makes sense, at least where individual objects are not involved, to ask whether the predicate (P) is contained in the subject (S) or not.

However modern logic recognizes many propositions which are not of the subject-predicate form e.g. $(\forall x)(\exists y)R(x,y)$, or, as Frege says, existential propositions such as $(\exists x)P(x)$. It is not surprising that Frege should make this criticism since he is one of the founders of modern logic. To overcome the difficulty, Frege proposes a new definition of analytic which corresponds closely to Kant's original intentions, but is adaptable to modern logic. We will consider this in the next chapter.

One of Kant's principal claims is that mathematical judgements are synthetic *a priori*. Kant first argues, (1783) Prolegomena, § 2, pp. 18-19, that:

"... properly mathematical propositions are always judgements *a priori*, and not empirical, because they carry with them necessity, which cannot be taken from experience."

He next argues that $7 + 5 = 12$ is synthetic because, by analysis of the concept of the unification of 7 and 5, we cannot obtain the concept 12. As he puts it (op. cit. § 2, pp. 19-20):

"The concept of twelve is in no way already thought by merely thinking this unification of seven and five, and though I analyse my concept of such a possible sum as long as I please, I shall never find the twelve in it. We have to go outside these concepts and with the help of the intuition which corresponds to one of them, our five fingers for instance or (as SEGNER does in his Arithmetic) five points, add to the concept of seven, unit by unit, the five given in intuition. Thus we really amplify our concept by this proposition 7 + 5 = 12, and add to the first concept a new one which was not thought in it. That is to say, arithmetical propositions are always synthetic, of which we shall be the more clearly aware if we take rather larger numbers. For it is then obvious that however we might turn and twist our concept, we could never find the sum by means of mere analysis of our concepts without seeking the aid of intuition."

The same applies to geometrical truths e.g. a straight line is the shortest distance between two points. 'Straight' is a qualitative concept and so we cannot, by analysis of concepts, obtain the quantitative fact that a straight line is the shortest distance between two points.

Kant does not of course deny that mathematics makes use of analytic truths e.g. (his example) the whole is greater than the part. His claim is rather that all significant mathematical truths are synthetic.

Having claimed that there exists synthetic *a priori* knowledge, Kant immediately raises the question of how such knowledge is possible. This is indeed a problem. If a judgement is synthetic i.e. not based on a mere analysis of concepts, it would appear to be 'about the world', and hence only knowable on the basis of experience. How then can we know about the world *a priori*? In order to explain Kant's answer in the case of mathematics, we must next introduce and explain the Kantian concept of *intuition*.

For Kant, intuition means much the same as 'sense-perception' or 'sensibility'. As he says, (1781) Critique of Pure Reason A51/B75 p. 93:

"Our nature is so constituted that our *intuition* can never be other than sensible; that is, it contains only the mode in which we are affected by objects. ... Without sensibility no object would be given to us, ..."

However (and this is where he differs from the empiricists), Kant holds that intuition contains not only a *matter*, originating from the thing which is being intuited, but also a *form* supplied by the human mind. Indeed he claims, (1781) Critique of Pure Reason, A22/B36, p. 67, that:

"... there are two pure forms of sensible intuition, serving a principles of *a priori* knowledge, namely, space and time."

Russell illustrates this theory, (1946) History of Western Philosophy, Book 3, Ch. XX Kant, p. 734, by the analogy of a man who, because he wears blue spectacles, sees everything blue. Similarly, according to Kant, we all wear spatio-temporal spectacles and so see everything

13

in space and time. Things-in-themselves, however, are outside space and time.

The connection between this and mathematics is provided by the following quotation, (1783) Prolegomena, § 7, p. 36:

> "But we find that all mathematical knowledge has this peculiarity, that it must first exhibit its concept *in intuition,* and do so *a priori,* in an intuition that is not empirical but pure; without this means mathematics cannot make a single step."

What Kant means here can best be seen by considering geometry. In order to prove e.g. Pythagoras' theorem, we must draw figures, or visualize such figures in our mind's eye. That is, we must, in Kant's terminology, exhibit the concepts (e.g. straight line, triangle, right angle, square, etc.) in intuition.

Similarly in arithmetic we have to proceed by counting — a process which takes time. However, as we saw in the earlier quotation, for arithmetic we need also spatial intuitions of such things as fingers or points. Apart from one brief reference to algebra, (1781) Critique of Pure Reason A734/B762 p. 590, Kant identifies mathematics with arithmetic and Euclidean geometry. He is thus able to link mathematics with intuitions of space and time in the above manner.

Kant's theory of space and time as forms of intuition, and of things-in-themselves outside space and time, seems strange and exotic. Yet granted his basic premises, it follows almost of necessity. Kant, like most thinkers before the discovery of non-Euclidean geometry, regarded Euclid's axioms as certain and necessary, and hence as not empirical. On this view, we know in advance that every object which we see and feel will obey these axioms. But how is such knowledge possible? It seems that it can only be explained on the assumption that spatial relations are supplied not by the object but by ourselves. To use Russell's analogy, we are certain that the world will obey Euclid's axioms only because we look at it through Euclidean spectacles.

Kant himself puts this line of thought as follows (1783) Prolegomena, § § 8 & 9, p. 38:

> "But how can *intuition* of the object precede the object itself?
>
> If our intuition had to be of such a nature that it represented things *as they are in themselves,* no intuition *a priori* would ever take place and intuition would be empirical everytime. ... There is thus only one way in which it is possible for my intuition to precede the reality of the object and take place as knowledge *a priori, namely if it contains nothing else than the form of sensibility which in me as subject precedes all real impressions through which I am affected by objects.* That objects of the senses can only be intuited in accordance with this form of sensibility is something that I can know *a priori."*

Perhaps the greatest blow to Kant's theory of mathematics was the discovery of non-Euclidean geometries and the proof that such geometries are consistent relative to Euclidean geometry. These results suggested that we do not after all know *a priori* the true geometry of space, but have to determine the matter *a posteriori*, on the basis of experience. However, Frege does not attack Kant on these grounds. Indeed he defends the Kantian view of geometry. His criticisms, as we shall see in the next chapter, are directed against Kant's theory of arithmetic.

Chapter 2.
Frege's Criticisms of Kant

Kant held that both geometry and arithmetic are synthetic *a priori*. Frege agreed that geometry was synthetic *a priori,* but argued that arithmetic was analytic rather than synthetic. As we have seen, Frege regarded Kant's definition of analytic as too narrow because it depended on a subject-predicate logic. He therefore begins his own treatment of the matter by giving a new definition of analytic. However he remarks in a footnote, (1884) Foundations of Arithmetic, § 3, p. 3e:

> "... I do not, of course, mean to assign a new sense to these terms, but only to state accurately what earlier writers, KANT in particular, have meant by them."

This comment seems to me fair. Frege's definition extends Kant's in a natural way which makes it appropriate to modern logic. The definition itself runs as follows (op. cit. § 3, p. 4e):

> "The problem becomes, in fact, that of finding the proof of the proposition, and of following it right back to the primitive truths. If, in carrying out this process, we come only on general logical laws and on definitions, then the truth is an analytic one, bearing in mind that we must take account also of all propositions upon which the admissibility of any of the definitions depends. If, however, it is impossible to give the proof without making use of truths which are not of a general logical nature, but belong to the sphere of some special science, then the proposition is a synthetic one."

Frege's main thesis is that the truths of arithmetic are analytic in this sense, and hence *a priori*. This must, at first sight, have seemed implausible since (i) arithmetic involves special entities — the natural numbers 0, 1, 2, 3, ..., n, ... — which look very different from anything which occurs in logic, and (ii) arithmetic involves special modes of reasoning — particularly the principle of mathematical, or complete, induction ($P(0) \wedge (\forall n)(P(n) \to P(n+1))$)) $\to (\forall n)P(n)$[1] — which appear to differ from ordinary logical reasoning. However, Frege hoped to overcome these objections by (i) defining number in terms

---

[1] For those unfamiliar with this principle, an account of it is given in Appendix II. On the Principle of Mathematical, or Complete, Induction.

of purely logical notions, and (ii) showing that mathematical induction can be reduced to ordinary logical inference. As he puts it in the introduction, (1884) Foundations of Arithmetic, p. IVe:

> "The present work will make it clear that even an inference like that from n to $n + 1$, which on the face of it is peculiar to mathematics, is based on the general laws of logic,..."

Most of Frege's book on the foundations of arithmetic is taken up with an attempt to establish that the truths of arithmetic are analytic. But as well as defending his own position, he attacks the Kantian view that these truths are synthetic. Kant's theory is based on the view that mathematical concepts must be exhibited in intuition. Thus, to take Kant's own example, the equation $7 + 5 = 12$ is verified as follows. The number five must be exhibited in intuition, and then added unit by unit to 7 to get 12. Kant's idea seems to be that, to check the equation, I must hold up the fingers of one hand and count from left to right 8, 9, 10, 11, 12. Of course the use of fingers here is inessential. A group of sticks, points or strokes on the blackboard etc. could be used instead. What is important is not the matter of the intuition but its form. On the other hand it is essential to Kant's theory that numbers be made preceptible in some way in order to verify the equations of arithmetic. It is just this which Frege challenges.

Frege's basic point is that Kant's view looks plausible for *small* numbers, but obviously breaks down for *large* ones. I shall call his argument here: *Frege's argument from large numbers,* since, as we shall see, it can be used in other contexts. To expound it, he changes Kant's example of $5 + 7 = 12$ to $135664 + 37863 = 173527$, and comments, (1884) Foundations of Arithmetic, § 5, p. 6e:

> "KANT thinks he can call on our intuition of fingers or points for support, thus running the risk of making these propositions appear to be empirical, contrary to his own expressed opnion; for whatever our intuition of 37863 fingers may be, it is at least certainly not pure. Moreover, the term "intuition" seems hardly appropriate, since even 10 fingers can, in different arrangements, give rise to very different intuitions. And have we, in fact, an intuition of 135664 fingers or points at all? If we had, and if we had another of 37863 fingers and a third of 173527 fingers, then the correctness of our formula, if it were unprovable, would have to be evident right away, at least as applying to fingers; but it is not."

Frege comes back to the same point later in the book (op. cit. § 89, p. 101e):

> "I must also protest against the generality of KANT's dictum: without sensibility no object would be given to us. Nought and one are objects which cannot be given to us in sensation. And even those who hold that the smaller numbers are intuitable, must at least concede that they cannot be given in intuition any of the numbers greater than $1000^{1000^{1000}}$, about which nevertheless we have plenty of information."

17

Frege's argument here seems to me quite correct, and of considerable importance. It is quite plausible to argue that arithmetical equations involving small numbers are based on "intuition" or "sense-perception" together with simple counting operations. But such an account breaks down for large numbers such as 173527. To handle such numbers we need to grasp a theoretical scheme (the decimal, or some similar, notation, together with the various rules for adding, subtracting etc. in that notation). Moreover this scheme goes beyond what is immediately and directly perceptible. We shall have occasion to examine further the implications of this in what follows.

It is interesting to note that in the passage from the Prolegomena quoted above in which Kant discusses his example $7 + 5 = 12$, he writes, (1783) Prolegomena, § 2, p. 19:

> "That is to say, arithmetical propositions are always synthetic, of which we shall be the more clearly aware if we take rather larger numbers."

Perhaps Frege followed Kant's suggestion, considered rather larger numbers, and was thus led to an argument against Kant's theory.

Although Frege attacks Kant on arithmetic, he defends him on geometry. Indeed he says, (1884) Foundations of Arithmetic, § 89, pp. 101e-102e:

> "I have no wish to incur the reproach of picking petty quarrels with a genius to whom we must all look up with grateful awe; I feel bound, therefore, to call attention also to the extent of my agreement with him, which far exceeds any disagreement. To touch only upon what is immediately relevant, I consider KANT did great service in drawing the distinction between synthetic and analytic judgements. In calling the truths of geometry synthetic and a priori, he revealed their true nature. And this is still worth repeating, since even to-day it is often not recognized. If KANT was wrong about arithmetic, that does not seriously detract, in my opinion, from the value of his work. His point was, that there are such things as synthetic judgements a priori; whether they are to be found in geometry only, or in arithmetic as well, is of less importance."

It might be asked how Frege reconciled this Kantianism with the existence (and consistency relative to Euclidean geometry) of systems of non-Euclidean Geometry. Such systems were in fact well-known to the mathematical community by the 1880's. Frege argues that, although systems of non-Euclidean geometry are logically possible, Euclidean geometry is the only one which agress with intuition. This is how he puts it, (1884) Foundations of Arithmetic, § 14, pp. 20e-21e:

> "... the truths of geometry govern all that is spatially intuitable, whether actual or product of our fancy. The wildest visions of delirium, the boldest inventions of legend and poetry, where animals speak and stars stand still, where men are turned to stone and trees turn into men, where the drowing haul themselves up out of swamps by their own topknots — all these remain, so long as they remain intuitable, still subject to the axioms of geometry. Conceptual thought alone can after a

fashion shake off this yoke, when it assumes, say, a space of four dimensions or positive curvature. To study such conceptions is not useless by any means; but it is to leave the ground of intuition entirely behind. If we do make use of intuition even here, as an aid, it is still the same old intuition of Euclidean space, the only space of which we have any picture. Only then the intuition is not taken at its face value, but as symbolic of something else; for example, we call straight or plane what we actually intuite as curved. For purposes of conceptual thought we can always assume the contrary of some one or other of the geometrical axioms, without involving ourselves in any self-contradictions when we proceed to our deductions, despite the conflict between our assumptions and our intuition."

It seems to me however that Kantian intuition i.e. direct perception is not sufficient to decide whether space is Euclidean, or whether it has a small positive or negative curvature. Such a decision can only be made on the basis of very delicate measurements using refined instruments. It is not a matter which can be determined *a priori*.

Frege's view on this question is indicative of his prejudice against empiricism, and in favour of *a priori* knowledge. Indeed Frege makes no secret of his opposition to empiricism. Regarding his philosophical views, he writes, (1884) Foundations of Arithmetic, Introduction, pp. X^e-XIe:

"Their reception by philosophers will be varied, depending on each philosopher's own position; but presumably those empiricists who recognize induction as the sole original process of inference (and even that as a process not actually of inference but of habituation) will like them least."

Moreover Frege criticizes Mill in very harsh terms, but before we come to this, we must first examine Mill's own views.

Chapter 3.
Mill's Theory of Mathematics

Mill's views on the philosophy of mathematics are set out in his (1843) A System of Logic Book II Chs 5 & 6, and Book III Ch. 24. Our main concern will be Mill's account of arithmetic, but, as in the case of Kant, we will consider his theory of geometry as well in order to form a better over-all picture of his approach to mathematics. Indeed, as the empiricist view of geometry is perhaps simpler and more plausible than the empiricist view of arithmetic, we will make it our starting point.

Mill thinks of geometry as Euclidean geometry, and consequently as a body of knowledge deduced from a few axioms and definitions. But while Kant held that Euclid's axioms were known *a priori*, Mill argues that they are founded on observation and experiment. As he says, (1843) A System of Logic. Book II. Ch. 5 § 4 pp. 151-2:

> "It remains to inquire, what is the ground of our belief in axioms — what is the evidence on which they rest? I answer, they are experimental truths; generalisations from observations. The proposition, Two straight lines cannot enclose a space — or, in other words, two straight lines which have once met do not meet again, but continue to diverge — is an induction from the evidence of our senses."

Mill believed that his empiricist theory of mathematics contradicted standard views on the subject, and would consequently be subject to severe attacks. As he says, op. cit. Book II. Ch. 5 § 4 p. 152:

> "This opinion runs counter to a scientific prejudice of long standing and great strength, and there is probably no proposition enunciated in this work for which a more unfavourable reception is to be expected."

Mill's gloomy expectations certainly proved justified — at least as far as Frege is concerned. However the opponent whom Mill himself had in mind was Whewell — the leading English Kantian of the day. Mill goes on immediately to expound the alternative views of Whewell, op. cit. Book II. Ch. 5. § 4 p. 152:

> "It is not necessary to show that the truths which we call axioms are originally *suggested* by observation, and that we should never have known that two straight lines cannot enclose a space of we had never seen a straight line: thus much being admitted by Dr. Whewell and by all, in recent times, who have taken his view of the subject. But they contend that it is not experience which *proves* the axiom; but

that its truth is perceived *à priori*, by the constitution of the mind itself, from the first moment when the meaning of the proposition is apprehended, and without any necessity for verifying it be repeated trials, as is requisite in the case of truths really ascertained by observation."

Much of Mill's discussion of geometry is taken up with criticizing the view that the truth of the axioms is based on some kind of *a priori* intuition. Before we come to this, however, let us examine briefly how he develops his positive views on geometry.

One of Mill's interesting claims is that the definitions of Euclidean geometry are, like the axioms, based on observation and experiment. Indeed the definitions are not merely verbal, but assert matters of fact. Thus the definition of circle asserts, according to Mill, that circles do actually exist in the empirical world. He speaks, (1843) A System of Logic. Book II. Ch. 6. § 2. p. 169, of:

"... definitions in the geometrical sense, not the logical; asserting not the meaning of a term only, but along with it an observed matter of fact. The proposition, "A circle is a figure bounded by a line which has all its points equally distant from a point within it," is called the definition of a circle; but the proposition from which so many consequences follow, and which is really a first principle in geometry, is, that figures answering to this description exist."

This opinion is somewhat doubtful. I can for example define a plane, regular, chiliagon (i.e. polygon with a thousand sides), and I can even from the definition prove theorems about such figures. For example it is an easy matter to calculate the magnitude of the (equal) angles of a plane regular chiliagon. Yet no figure answering to this description exists in the sense required by Mill i.e. has been drawn on paper, or constructed in some other way.

There is also a difficulty even in the case of a simple figure such as a circle. Now, without doubt, numerous circles have been drawn and constructed in various ways. However, it could be objected that such circles are not really circles in the full geometrical sense, because, for example, their various radii will never be exactly equal and so on. More generally against Mill's view that the axioms of geometry are verified empirically, it could be objected that these axioms are really contradicted by experience, since we never in practice meet with infinitely thin lines, points without extensions, and the other entities dealt with by Euclid.

Mill himself considers objections of this kind, and he replies that the points, lines and circles of Euclid are only approximations to the points, lines and circles which exist in the material world. However this approximation is generally speaking good enough for practical purposes. As he says (1843) A System of Logic. Book II. Ch. 5 § 1 p. 148:

> "The correctness of those generalisations, *as* generalisations, is without a flaw: the equality of all the radii of a circle is true of all circles, so far as it is true of any one: but it is not exactly true of any circle; it is only nearly true; so nearly that no error of any importance in practice will be incurred by feigning it to be exactly true."

Moreover, as Mill goes on to say, corrections for the finite thickness of lines etc. can always be introduced, should this be necessary, in a particular case. He also argues, op. cit. Book II. Ch. 5. § 4. p. 153 Footnote, that we can obtain inductive support for Euclid's axioms by observing ever thinner lines and extrapolating.

Mill seems to me to be on strong ground here. In many sciences which are generally agreed to be empirical rather than *a priori*, we find that use is made of entities which agree only approximately and not exactly with what is observed. An obvious example comes from the theory of gases. Here physicists consider the ideal gas and its laws. No observed gas is ever ideal, but calculations about ideal gases are not useless because certain actual gases under certain circumstances do approximate in their behaviour to that of an ideal gas. Moreover where deviations from ideal gas behaviour do occur, corrections can be introduced, and the deviations at least partly explained.

Mill does not deny that we believe strongly in the truth of the axioms of geometry, but he claims that this strong belief can be explained by the fact that there is very strong observational and experimental evidence for these axioms. We do not need to appeal to any *a priori* intuitions to convince ourselves of the truth of geometry. As he says, (1843) A System of Logic. Book II. Ch. 5. § 4. p. 152:

> "... the truth of the axiom, Two straight lines cannot enclose a space ... is ... evident from experience. ... the axiom ... receives confirmation in almost every instant of our lives, since we cannot look at any two straight lines which intersect one another without seeing that from that point they continue to diverge more and more. Experimental proof crowds in upon us in such endless profusion, and without one instance in which there can be even a suspicion of an exception to the rule, that we should soon have stronger ground for believing the axiom, even as an experimental truth, than we have for almost any of the general truths which we confessedly learn from the evidence of our senses."

Since there is such strong empirical evidence for the truth of the axioms of geometry (or more strictly for claiming that they hold to a high degree of approximation), there is no need to postulate *a priori* intuitions. Mill thinks that the burden of proof is on his opponents to demonstrate that such intuitions exist, and that their attempts to do so are largely unsuccessful.

Mill is prepared to concede that we can verify the axioms of geometry by what Whewell calls "imaginary looking", and what

Kant himself would probably describe as "exhibiting the concept in a (pure) intuition". However he argues that this is only so because, (1843) A System of Logic. Book II. Ch. 5. § 5. p. 154:

> "... the imaginary lines exactly resemble real ones, ..."

Moreover Mill argues, following a certain Professor Bain, that, op. cit. Book II. Ch. 5. § 5. p. 155:

> "The psychological reason why axioms, and indeed many propositions not ordinarily classed as such, may be learnt from the idea only, without referring to the fact, is that in the process of acquiring the idea we have learnt the fact. The proposition is assented to as soon as the terms are understood, because in learning to understand the terms we have acquired the experience which proves the proposition to be true."

Mill now goes on to criticize the Kantian view that some propositions have, (1783) Prolegomena § 2 pp. 18-19, a "necessity, which cannot be taken from experience." Following Whewell, Mill quite reasonably takes a necessary proposition to be one whose negation is not only false but inconceivable, and he comments, (1843) A System of Logic. Book II. Ch. 5 § 6 pp. 156-7:

> "This, therefore, is the principle asserted: that propositions, the negation of which is inconceivable, or in other words, which we cannot figure to ourselves as being false, must rest on evidence of a higher and more cogent description than any which experience can afford."

Mill replies along Humean lines that the alleged necessity of these propositions is a mere psychological illusion caused by the fact either that the propositions are based on very familiar experience, or perhaps simply by the fact that the propositions themselves are very familiar and have become an established part of our thinking. As Mill himself says, (1843) A System of Logic. Book II. Ch. 5. § 5. p. 157:

> "Now I cannot but wonder that so much stress should be laid on the circumstance of inconceivableness, when there is such ample experience to show that our capacity or incapacity of conceiving a thing has very little to so with the possibility of the thing in itself, ... There is no more generally acknowledged fact in human nature than the extreme difficulty at first felt in conceiving anything as possible which is in contradiction to long-established and familiar experience, or even to old familiar habits of thought."

Mill argues for this thesis by citing a number of examples, from the history of science, of propositions which were once thought to be inconceivable, but are now held to be true. For example it was once thought to be inconceivable that men should live at the antipodes since they would fall off. The Cartesians, including Leibniz, held that gravitational attraction at a distance was inconceivable. The opponents of Copernicus regarded the motion of the earth as inconceivable, and so on.

Mill discusses attitudes to Newton's first law of motion (Every body continues in its state of rest, or of uniform motion in a right line, unless it is compelled to change that state by forces impressed upon it) as showing that familiarity produces the illusion of necessity. At first the law seemed to people obviously false, since it appears to be a common fact of experience that motions, once acquired, tend to diminish gradually and then cease. However, no sooner had the law become part of established science than the Kantians claimed it to be *a priori* necessary.

Mill ironically cites Whewell as a remarkable instance of the psychological law that familiarity creates the illusion of necessity. Whewell, Mill argues, being such a learned scientist, had made himself so familiar with the laws of chemical composition that he had come to regard them as *a priori* necessary while their original discoverer was still living, even though the laws had only been established by laborious and exact experiments.

This polemic of Mill's against Whewell and the Kantians is not only sparkling in style, but compelling in content. The verdict of history is undoubtedly that many propositions once supposed to be necessarily true were later shown to be false. Thus the alleged necessity of any proposition should always be regarded with suspicion as possibly a psychological illusion.

Mill's empiricist view of geometry was later to be remarkably strengthened by the discovery of non-Euclidean geometry, and its application in general relativity. Once again propositions which had been held to be *a priori* necessary (Euclid's axioms) were shown to be false, or, perhaps better, to hold only approximately.

When Mill wrote in 1843, he could in principle have known of non-Euclidean geometry, since Bolyai and Lobachevsky had already published their systems of hyperbolic geometry, and Lobachevsky had even produced versions of his theory in both French and German. However Mill clearly did not know of their work, since he speaks of the axioms of Euclidean geometry as, (1843) A System of Logic. Book II. Ch. 5. § 6 p. 160, convictions:

"of the conclusiveness of which, from the earliest records of human thought, no sceptic has suggested even a momentary doubt..."

Mill's ignorance of non-Euclidean geometry is not surprising since it did not become widely known in Western Europe until the late 1860's. Thus non-Euclidean geometry provided striking independent evidence for a view which Mill originally advocated on other grounds.

Mill's treatment of arithmetic is along the same lines as his treatment of geometry. He begins by remarking, (1843) A System of Logic. Book II. Ch. 6. § 1 p. 166:

"It is harder to believe of the doctrines of this science than of any other, ... that they are not truths *à priori*, but experimental truths..."

Nonetheless Mill claims that this is indeed the case. For example the proposition '2 + 2 = 4' is based on a number of facts which may be verified by experiment and observation such as the following. If I count out 2 apples and put them into an empty box, and count out 2 more apples and put them into the box, then, if I count the apples in the box, I will arrive at the figure four. Again if I count the apples in the box in different orders, I will still reach the number four, and the same will apply if I arrange the apples into a different shape and then recount them. The laws of arithmetic are simply inductive generalisations from observed facts such as these. This is how Mill himself puts it, (1843) A System of Logic. Book II. Ch. 6. § 2. pp. 168-9:

"Three pebbles in two separate parcels, and three pebbles in one parcel, do not make the same impression on our senses; and the assertion that the very same pebbles may by an alteration of place and arrangement be made to produce either the one set of sensations or the other, though a very familiar proposition, is not an identical one. It is a truth known to us by early and constant experience — an inductive truth; and such truths are the foundation of the science of Numbers. The fundamental truths of that science all rest on the evidence of sense; they are proved by showing to our eyes and our fingers that any given number of objects, ten balls, for example, may by separation and rearrangement exhibit to our senses all the different sets of numbers the sum of which is equal to ten. All the improved methods of teaching arithmetic to children proceed on a knowledge of this fact. All who wish to carry the child's *mind* along with them in learning arithmetic; all who wish to teach numbers, and not mere ciphers — now teach it through the evidence of the senses, in the manner we have described."

Mill remarks that we can regard the equation 3 = 2 + 1 as a definition of the number three, but such definitions, as in the case of geometry, really assert facts, op. cit. Book II. Ch. 6. § 2. p. 169:

"And thus we may call "Three is two and one" a definition of three; but the calculations which depend on that proposition do not follow from the definition itself, but from an arithmetical theorem presupposed in it, namely, that collections of objects exist, which while they impress the senses thus, $^oo^o$, may be separated into two parts, thus, oo o. This proposition being granted, we term all such parcels Threes, after which the enunciation of the above-mentioned physical fact will serve also for a definition of the word Three."

In the case of geometry, Mill argued that the axioms and theorems are not exactly true, but hold only approximately of empirical lines, points, circles etc. Indeed he seems to regard this as being a characteristic feature of laws founded on experience. Such laws either hold only approximately, or admit of exceptions which are tacitly excluded. As an example of the latter situation, we might consider the law that man is a biped. Now strictly speaking this law is contradicted by anyone who has had a leg amputated, or by the occasional freak cases of people born with three legs. However, such exceptions are tacitly excluded when the law is stated. Indeed this convention of tacit exclusion is employed in any textbook of biology which describes the characteristics of animal and plant species.

Now if the laws of arithmetic are founded on observation and experiment, they too should, according to Mill, hold only approximately or admit of exceptions. Indeed Mill tries to give examples to show that this is really the case. He argues that arithmetical truths are not exactly correct in some cases, because the empirical units involved may not be exactly equal, (1843) A System of Logic. Book II. Ch. 6. § 3. p. 170:

> "How can we know that one pound and one pound make two pounds, if one of the pounds may be troy, and the other avoirdupois? They may not make two pounds of either, or of any weight."

This example is perhaps not a very happy one, since it could be argued that, when we apply equations like $1 + 1 = 2$ to weighing substances, there is a tacit convention that the same system of units is used throughout the equation. However there are a number of familiar examples which seem better adapted to making Mill's point.

Suppose we put two rabbits into an empty enclosure, and then another two. We may find after a lapse of time that there are not four rabbits in the enclosure, but five, six or seven — even though no rabbits have been added or removed. Again if we add one drop of water to another drop, we may find not two drops of water but one larger drop. Such examples are sometimes cited to show that the truths of arithmetic are not empirical. The argument is that if equations like $2 + 2 = 4$ or $1 + 1 = 2$ were empirical generalisations, we would regard them as refuted by the rabbits or water drops; but we do not in fact takes these counter-examples as refuting the equations, and this, it is claimed, shows that the equations are not empirical laws.

It seems to me that the opposite conclusion could be drawn from these examples. The laws of arithmetic are empirical generalisations, but hold only under certain conditions, namely that the objects counted should not reproduce or coalesce etc. As with other empirical laws, these limiting conditions are tacitly assumed without being explicitly stated.

This concludes my account of Mill's view of arithmetic. Let us now turn to Frege's criticisms.

Chapter 4.
Frege's Criticisms of Mill

We have seen that Frege regarded Kant as, (1884) Foundations of Arithmetic. § 89. p. 101e:

> " a genius to whom we must all look up with grateful awe";

and Frege's disagreement with Kant is only partial. Frege's attitude towards Mill is very different. He appears to regard Mill as a blockhead, and always refers to him in a contemptuous polemical tone. Thus, typically, Frege writes of Mill, op. cit. § 7. p. 9e:

> "... but this spark of sound sense is no sooner lit than extinguished."

and again, op cit. Introduction. p. VIIe:

> "What, then, are we to say of those who, instead of advancing this work where it is not yet completed, despise it, and betake themselves to the nursery, ... there to discover, like JOHN STUART MILL, some gingerbread or pebble arithmetic! It remains only to ascribe to the flavour of the bread some special meaning for the concept of number."

The tone of these remarks is indicative of Frege's hostility to empiricism — a hostility which we have already remarked on.

Mill, it will be remembered, held that we can define numbers as 2 = 1 + 1, 3 = 2 + 1, 4 = 3 + 1, ... − this is his "spark of sound sense". However, he also held that these definitions depend on general empiral facts about the world which are established by experience and induction. These are facts about the existence of discrete objects, about possible arrangements of such objects, etc. This empiricist doctrine is what, for Frege, extinguishes the spark of sound sense.

Let us now summarize Frege's arguments against Mill. First of all Frege uses his favourite device of considering large instead of small numbers. We may think of 2 + 1 = 3 as established by sense experience, but what of: 999,999 + 1 = 1,000,000? (1884) Foundations of Arithmetic § 7 p. 9e:

> "But what in the world can be the observed fact, or the physical fact (to use another of MILL's expressions), which is asserted in the definition of the number 777864?"

and again, op. cit. § 7. pp. 10e-11e:

> "On MILL's view we could actually not put 1,000,000 = 999,999 + 1 unless we had observed a collection of things split up in precisely this peculiar way, ..."

Hereagain it must be admitted that Frege has a good point. It is quite plausible to argue that equations involving small numbers, such as 2 + 1 = 3 or 2 + 2 = 4, can be verified experimentally by counting pebbles or apples; but this account breaks down for equations involving large numbers such as Frege's 135664 + 37863 = 173527.

We argued in chapter 2 that Frege's argument from large numbers is fatal to the Kantian view that arithmetical truths are founded on intuition. It seems to me, however, that the argument is a much less decisive criticism of Mill's empiricist views. It may make some modification of these views necessary, but does not refute them completely.

The point here is that many mathematical sciences, which are generally admitted to be founded on observation and experiment, nonetheless contain equations which cannot be directly verified by sense experience. These equations are believed because they have consequences which can be checked and verified experimentally. An obvious example here is Mawell's equations. We cannot directly verify that .e.g. curl $\mathbf{H} = 4\Pi\mathbf{j} + \frac{1}{c}\frac{\partial \mathbf{D}}{\partial t}$ in any simple way. But, nonetheless, our belief in Maxwell's equations is ultimately founded in observations of electromagnetic phenomena. Similarly the equations of decimal arithmetic involving large numbers are believed because of the (indirect) confirmation they receive in a multitude of observations, experiments, and practical applications. This, however, is perhaps a development of Mill's position. Let us therefore return to Mill himself.

Frege rejects Mill's account for small numbers as well as for large. He says, (1884) Foundations of Arithmetic § 7 p. 9e:

> "Of all the whole wealth of physical facts in his apocalypse, MILL names for us only a solitary one, the one which he holds is asserted in the definition of the number 3. It consists, according to him, in this, that collections of objects exist, which while they impress the senses thus, o_oo, may be separated into two parts, thus, oo o. What a mercy, then, that not everything in the world is nailed down; for if it were, we should not be able to bring off this separation and 2 + 1 would not be 3! What a pity that MILL did not also illustrate the physical facts underlying the numbers 0 and 1!"

Mill could perhaps reply here that of everything in the world was nailed down, human life would become impossible, and, even if intelligent beings could exist under these circumstances, it is not likely that they would have arithmetic. Not is it clear why Frege thinks that there is a special difficulty regarding the numbers 0 and 1.

Suppose we accept that the equation $2 + 2 = 4$ can be verified empirically by counting apples into a box. We could perform an exactly similar verification for $0 + 1 = 1$. It would go like this. Take an empty box, and place one apple in it. It will then be found that there is one apple in the box. It is certainly true that the facts here are rather trivial, but, according to the observations of Piaget and other child pyschologists, such facts do at first surprise and interest young babies.

Frege's next argument is that, on Mill's view, one could not have a number of non-physical objects. Indeed the passage just quoted continues, op. cit. § 7. pp. 9e-10e:

> " "This proposition being granted," MILL goes on, "we term all such parcels Threes." From this we can see that it is really incorrect to speak of three strokes when the clock strikes three, or to call sweet, sour and bitter three sensations of taste; and equally unwarrantable is the expression "three methods of solving an equation." For none of these is a parcel which ever impresses the senses thus, $o_o o$."

Here Frege is being somewhat unfair, for Mill does explicitly state that arithmetic applies to objects of any kind, (1843) A System of Logic. Book II. Ch. 6. § 2. p. 167:

> "Propositions ... concerning numbers have the remarkable peculiarity that they are propositions concerning all things whatever; ... . That half of four is two, must be true whatever the word four represents, whether four hours, four miles, or four pounds weight."

Nor is this difficult to accomodate on Mill's empiricist position. $2 + 1 = 3$ may be verified first for apples, sticks and pebbles, but we can then learn by further experience that it applies to strokes of the clock etc.

Frege next accuses Mill of confusing arithmetical propositions with their applications. For Frege, 2 apples and 2 apples make 4 apples is an application of $2 + 2 = 4$, but the two kinds of proposition are distinct, (1884) Foundations of Arithmetic § 9 p. 13e:

> "MILL understands the symbol + in such a way that it will serve to express the relation between the parts of a physical body or of a heap and the whole body or heap; but such is not the sense of that symbol. That if we pour 2 unit volumes of liquid into 5 unit volumes of liquid we shall have 7 unit volumes of liquid, is not the meaning of the proposition $5 + 2 = 7$, but an application of it, which only holds good provided that no alteration of the volume occurs as a result, say, of some chemical reaction. MILL always confuses the applications that can be made of an arithmetical proposition, which often are physical and do presuppose observed facts, with the pure mathematical proposition itself."

Mill might reply that you cannot separate the pure mathematical proposition from its applications, since the meaning of the former depends on the latter. Such a view would gain support from the theory of meaning in Wittgenstein's later philosophy. According to

Wittgenstein, the meaning of a word is given by its use in various social activities. Thus to understand the meaning of the symbols 2, 5, 7 etc., someone would have to know how they are used in everyday social life. For example he would have to know how they are employed in counting out bricks on the building site, to use Wittgenstein's favourite example (c.f. (1953) Philosophical Investigations § 8 p. 5e). This point of view is also supported by the fact that we could not teach children the meaning of number words without at the same time teaching how they are used in the typical applications of ordinary life.

Frege has one last interesting argument against founding arithmetic on experience and induction. He argues that induction depends on probability and hence on arithmetic, rather than vice versa, (1884) Foundations of Arithmetic § 10, pp. 16e-17e:

> "Induction must base itself on the theory of probability, since it can never render a proposition more than probable. But how probability theory could possibly be developed without presupposing arithmetical laws is beyond comprehension."

Various answers could be made to this argument depending on the attitude adopted to confirmation theory. Here I shall just state my own view.

There is no doubt that, in both science and everyday life, people do speak of evidence confirming, supporting, rendering probable, corroborating, etc. a theory or a prediction. This relation of confirmation is certainly needed in empirical reasoning. Now the Bayesians hold that confirmation can be rendered quantitative i.e. measured by a real number (which may in the subjective version vary from individual to individual). They further hold that these numbers satisfy the axioms of probability. This seems to be the view of Frege.

To me, however, it seems most unlikely that confirmation will ever be rendered quantitative in a satisfactory manner. We can certainly make qualitative judgements of confirmation e.g. the evidence e supports theory T_1 better than theory T_2, but I doubt whether it will ever be possible to derive non-arbitrary values for the degree of such support e.g. e supports T_1 to degree 0.63, and T_2 to degree 0.76. Moreover such measures of support, even if they were possible, would not seem to be of much use. Science and technology have progressed very satisfactorily to date without a quantitative concept of confirmation, and it is highly doubtful whether, and in what way, such a concept would improve scientific and technological practice. But it confirmation remains a qualitative concept, it does not presuppose arithmetic, and Frege's argument falls to the ground.

Chapter 5.
The content of a statement of number is an assertion about a concept

Let us now turn to an exposition of Frege's own positive views regarding the foundations of artihmetic. We have already pointed out that, in order to establish his logicist thesis, Frege has to give a definition of number in terms of purely logical notions, and to show that all arithmetical reasoning, particularly reasoning from n to $n + 1$, or the principle of mathematical induction, can be reduced to logical inference. We will discuss Frege's attempts to do these things in chapter 7, but, before coming on to this, we will give an account of two doctrines of Frege's which have considerable interest irrespective of whether the logicist thesis is correct or not. One of these concerns the question of whether and in what sense numbers may be said to exist. We will deal with this in Ch. 6. The other will be the subject of the present chapter. Frege sums it up as the claim that the content of a statement of number is as assertion about a concept. We will describe it, rather more briefly, as the claim that numbers are properties of concepts and not of external things. This formulation is not strictly correct according to Frege for reasons which we shall explain in due course, but it will nonetheless serve us as a useful shorthand.

Consider then an expression like: 'four thoroughbred horses'. (This is Frege's own example c.f. (1884) Foundations of Arithmetic § 52 p. 64e). Now here 'thoroughbred' is a property of the horses, and the grammatical construction of the phrase suggests that 'four' too might be a property of the horses, and that in general number might be a property of external things. However Frege rejects this conclusion. Number, he argues, depends not just on the things considered, but on how they are considered. For example, (1884) Foundations of Arithmetic § 25 p. 33e:

"*One* pair of boots may be the same visible and tangible phenomenon as *two* boots."

and again, (1884) Foundations of Arithmeticc § 22 p. 28e:

> "If I give someone a stone with the words: Find the weight of this, I have given him precisely the object he is to investigate. But if I place a pile of playing cards in his hands with the words: Find the Number of these, this does not tell him whether I wish to know the number of cards, or of complete packs of cards, or even say of honour cards at skat[1]."

What Frege says here is a little doubtful, since it would be most natural to assume, in the circumstances he describes, that the number of cards was being requested. However, he is undoubtedly right that one could instead ask for the number of complete packs of cards, or of honour cards at skat.

From examples such as the boots and the cards, Frege draws the conclusion that number is not a property of external things, but of *concepts*. (1884) Foundations of Arithmetic § 46, p. 59e:

> "... the content of a statement of number is an assertion about a concept. This is perhaps clearest with the number 0. If I say "Venus has 0 moons", there simply does not exist any moon or agglomeration of moons for anything to be asserted of; but what happens is that a property is assigned to the *concept* "moon of Venus", namely that of including nothing under it. If I say "the King's carriage is drawn by four horses", then I assign the number four to the concept "horse that draws the King's carriage"."

In a digression from his main theme, Frege uses this doctrine to criticize the ontological argument for the existence of God. This argument may be stated as follows. Consider a being who has all the perfections i.e. all the admirable qualities such as goodness, truth, wisdom, justice, etc. Such a being is possible since no two perfections contradict each other. Therefore such a being must actually exist, since existence is a perfection.

Frege replies that existence, like number, is a property of concepts. Indeed any existence statement is an assertion that one or more things fall under a particular concept. To say that horses exist is to say that at least one object falls under the concept 'horse'. On the other hand perfections are properties of things not concepts. Thus existence cannot be a perfection, and the ontological argument fails. This is how Frege himself puts it, (1884) Foundations of Arithmetic, § 53, p. 65e:

> "In this respect existence is analogous to number. Affirmation of existence is in fact nothing but denial of the number nought. Because existence is a property of concepts the ontological argument for the existence of God breaks down."

It will be noted that Frege does here describe existence as a property of concepts. In the case of number he holds that this way of taking is, strictly speaking, illegitimate. This is because, in his philosophy of

---

[1] Skat: a three-handed card-game much played in Germany. (Oxford English Dictionary).

logic, he draws a strict distinction between 'concepts' and 'objects'. A concept has a 'hole' or 'holes' in it e.g. ... is a horse, or,... is to the left of ..., where the dots indicate the holes. An object is something which fills these holes. Now as we shall see in the next two chapters, numbers are, for Frege, objects and hence not concepts or properties. Thus instead of saying that number is a property of concepts, we should use the circumlocution: 'the content of a statement of number is an assertion about a concept'. We will however continue to use the shorter, though, on Frege's view somewhat less accurate, mode of expression.

Frege's arguments on this point are interesting, but it is possible to draw from them a conclusion somewhat different from his own. To see this let us consider again his example of the pile of playing cards. Suppose we have a particular such pile, call it Π. We can then associate with Π several different numbers viz. the number of cards in Π, the number of complete packs in Π, the number of honour cards at skat in Π etc. Frege concludes that we cannot associate numbers with particular physical objects such as Π, but only with concepts. In this instance we would have the following concepts:

$\text{Con}_C (...) =_{def} ...$ is a card in Π

$\text{Con}_P (...) =_{def} ...$ is a complete pack in Π

$\text{Con}_H (...) =_{def} ...$ is an honour card at skat in Π

However there is another possibility. Instead of associating numbers with concepts, we could associate them with sets. In this case we would have the following sets:

$\text{Set}_C =_{def}$ the set of cards in Π

$\text{Set}_P =_{def}$ the set of complete packs in Π

$\text{Set}_H =_{def}$ the set of honour cards at skat in Π.

At first sight it might seem that this is a completely trivial alteration of Frege's suggestion since the sets here are just the extensions of the corresponding concepts. So, it might be argued, it is a matter of indifference whether we consider sets or concepts. Indeed, in the usual notation, we have that, if Con (...) is an arbitrary concept, the extension of $\text{Con}(...) =$ the set of x such that $\text{Con}(x) = \hat{x} \, \text{Con}(x)$; and in this case

$\text{Set}_C = \hat{x} \, \text{Con}_C (x)$

$\text{Set}_P = \hat{x} \, \text{Con}_P (x)$

$\text{Set}_H = \hat{x} \, \text{Con}_H (x)$

However, as we shall see in chapter 12, it was precisely the transition from a concept to its extension which gave rise to Russell's paradox. This transition cannot therefore be considered simple and unproblematic, but, on the contrary, obscure and difficult. Correspondingly we cannot take the two suggestions: (1) numbers are properties of sets, and (2) numbers are properties of concepts to be straightforwardly equivalent. One of the two suggestions must be preferred, and I shall next argue against Frege that (1) is better than (2).

Before stating my argument, it will be as well to make a remark about terminology. In what follows, I shall use the terms: "set", "class", "collection", "aggregate", etc. as interchangeable. Now for certain purposes e.g. NBG set theory,[1] it is necessary to make distinctions between some of these terms — in the case cited between "set" and "class". But, for our own purposes, distinctions of this sort will not be necessary.

Our thesis then is that numbers are better considered as properties of sets (or classes or collections or etc.) than as properties of concepts. It is interesting to note that Frege considers and explicitly rejects this view, (1884) Foundations of Arithmetic § 45, p. 58ᵉ:

> "The terms "multitude", "set" and "plurality" are unsuitable, owing to their vagueness, for use in defining number."

As we shall see, he attacks the set-theoretic approach further when crticizing Dedekind and Schröder. I myself have two reasons for preferring sets to concepts. The first is that sets, classes, or collections do occur naturally as part of the material world, whereas concepts are often introduced secondarily merely as ways of describing these naturally occurring aggregates. The second is that a treatment based on sets is mathematically much simpler than one based on concepts. Here I believe is a case in which simplicity is the sign of truth.

Let us begin with the more metaphysical line of thought. Material reality consists of material things (whether stones, plants, animals or human beings) which stand in definite relationships to each other. These relationships have as much reality as the material things themselves. Thus, for example, a plant growing in the ground, and an insect flying through the air, may, at first sight, seem two distinct and separate material things, but yet in reality they are strongly interconnected. Many flowers cannot reproduce without being pollinated by insects, while many insects cannot survive without obtaining nectar from flowers. Note also that these relationships have nothing to do with human consciousness since they existed long before the appearance of man.

Now, as a result of these objective relationships, some things are

---

[1] The form of set theory developed by von Neumann, Bernays, and Gödel.

bound together to form naturally occurring aggregates. An obvious example of such an aggregate is a colony of bees. This consists of one queen, 40,000 or 50,000 workers, and a few hundred drones. Well-known relationships exist between these various types of bee. The queen lays the eggs. The workers collect nectar, turn it into honey, and tend the young. The drones have the sole function of fertilizing the queen, and are, with the approach of winter, expelled. (c.f. R. Chauvin, (1963) Animal Societies Ch. 1). Once again these relationships have nothing to do with human consciousness, since, as Chauvin writes, op. cit. Introduction p. 11:

> "Ants and bees were already in existence 40 million years ago at least, and scarcely differ from those we know today. ... And *Homo sapiens* has hardly 150,000 years of existence..."

Other examples of naturally occurring sets would be: the leaves of a tree, the trees of a wood, the planets of the solar system, the cells of a living organism, etc. Now, if the matter is considered carefully, it will be found that numbers are nearly always used in connection with such naturally occurring aggregates, where I here use 'natural' or 'material' in a broad sense to include human beings and their artefacts: that is, I consider man as part of the natural or material world. True, we can form bizarre sets such as that consisting of the handkerchief in my pocket, the President of the United States, and the Eiffel Tower; and there is no difficulty in seeing that this set has three members. However, such cases are artificial examples, and the usual applications of arithmetic deal with naturally occurring collections.

Now, whenever we have a naturally occurring set S, we can always introduce a corresponding concept (Con(x) say) such that x falls under the concept if and only if it belongs to the set. This concept is a way of describing the members of the given set. However, if we take the concept as being more fundamental than the set itself, we can easily be led into error, for it then looks as if what binds the objects together to form the set S is that they all possess the property expressed by Con(x), whereas, in reality, the members of S are often bound together by a complicated series of relationships about which Con(x) tells us nothing. We can illustrate this error by considering the following passage from Frege, (1884) Foundations of Arithmetic § 48 p. 61e:

> "The concept has a power of collecting together far superior to the unifying power of synthetic apperception. By means of the latter it would not be possible to join the inhabitants of Germany together into a whole; but we can certainly bring them all under the concept "inhabitant of Germany" and number them."

This passage seems to me to contain both correct and incorrect points. It is certainly correct that the millions of inhabitants of Germany cannot be viewed together in a Kantian intuition. This is Frege's argument from large numbers again. On the other hand it is not true that what joins the inhabitants of Germany together is the unifying power of a concept. In Frege's day the inhabitants of Germany did form a naturally occurring collection — a nation state. This state had indeed only recently come into existence, and it is hardly correct to say that it was formed by purely conceptual means. Indeed J. A. S. Granville writes, (1976) Europe Reshaped 1848-1878 Ch. XV. The Unification of Germany p. 303:

"Later patriotic myths have obscured the fact that Prussia in 1866 made war, not only on Austria, but also on the majority of the German states."

The inhabitants of a modern nation state are connected together by a complicated series of relationships. These relationships, though not physical in any direct sense, are nonetheless very real, since, for example, they enable one nation state to wage war against another, as Germany did against France in 1870.

Exactly the same argument applies to Frege's other example of the horses drawing the King's carriage. These horses certainly form a naturally occurring aggregate. They are not, however, bound together by the concept: "horse that draws the King's carriage", but rather by a system of harnessing.

The number "0" poses no difficulty for this doctrine, since it can be associated with the empty set. However, 0 and the empty set are in a certain sense degenerate or limiting cases. Consider Frege's example: "Venus has 0 moons". The idea of considering the empty set of moons of Venus surely arose in the following way. The naturally occurring non-empty sets of moons of Jupiter and of the Earth were observed. It was then an obvious step to consider: 'the set of moons of x' where x is an arbitrary planet, and in this way sets were defined which turned out to be empty. Such empty sets would surely never have been thought of had there not been non-empty sets of the same general type. Thus empty sets are, in a sense, parasitic on non-empty sets.

Admittedly the notion of empty set did involve some difficulties which, as we shall see, troubled those of Frege's contemporaries (Dedekind and Schröder) who advocated the set theoretic approach. We shall consider these difficulties when we come on (in chapter 9) to examine Dedekind's views in detail. Our conclusion will be that the problems involved can be overcome while still regarding the notion of set as more basic than that of concept.

Frege himself drew the opposite conclusion. In his interesting (1895) A Critical Elucidation of Some Points in E. Schröder;s *Vorlesungen über die Algebra der Logik*, he considers the problems which the empty set creates for Schröder's logic. Frege concludes that these difficulties can only be satisfactorily resolved by taking concepts as basic, and regarding sets as extensions of concepts. He writes (op. cit. pp. 103-104):

"... what else is there to constitute a class, if we ignore the concepts, the common properties! ... Only because classes are determined by the properties that individuals in them are to have, ... does it become possible to express thoughts in general by stating relations between classes; only so do we get a logic."

My answer to this has already been given. A class may be constituted not by its members all having a common property, but by a complicated system of relationships which hold between these members, or perhaps between these members and other parts of the material world.

Let us now turn to the mathematical considerations which favour regarding 'set' as more basic than 'concept', as far as the foundations of arithmetic are concerned. As we shall argue later on (Chs 8 and 11), regarding number as a property of concepts leads naturally to the development of 'higher-order logic', such as we find in Frege himself and then later in Russell's ramified theory of types. Regarding number as a property of sets, on the other hand, leads to the set-theoretic approach to arithmetic such as we find first in Dedekind and then in the axiomatic set theory of Zemelo, Fraenkel, and others. Now what is beyond doubt is that Zermelo-Fraenkel set theory turned out to be mathematically simpler than *Principia Mathematica*, and that it is much easier to develop arithmetic within set theory than within type theory and higher-order logic. This seems to me to provide additional support for the claim that numbers are better considered as properties of sets than as properties of concepts.

Chapter 6.
Frege's Platonism

Anyone reflecting about numbers and arithmetic will soon find himself or herself faced with what could be called the Platonic problem — the question, that is, of whether numbers exist, and, if so, in what sense. Simple considerations suggest that we should recognize the existence of numbers. For example, anyone, if asked whether there are numbers between 4 and 7, would reply: 'Yes there are, namely 5 and 6'. But, if there are numbers between 4 and 7, it follows logically that there are numbers i.e. that numbers exist. On the other hand, admitting that numbers exist often gives rise to a feeling of unease, since numbers appear to be curious shadowy entities very different from familiar material things such as tables and chairs, cats and dogs, other people etc.

We can distinguish three views regarding the existence of numbers. The first is that numbers do exist and are the subjective mental constructions of individuals. This view is sharply citicized by Frege, but was later adopted by Brouwer. The second is that numbers do exist, but are objective rather than subjective or psychological. This view could be called Platonism, and is supported by Frege. The third is that numbers do not after all exist, since talk of numbers is always reducible to talk of material things. This view could be called reductionism. It is perhaps most easily explained by considering statements about the average Englishman.

Let us suppose it to be true that p, where p ≡ the average Englishman has 2½ children. It is reasonable to suppose that every true statement such as p apparently about the average Englishman is reducible to a logically equivalent statement in which no mention is made of the average Englishman. For example we can contruct a statement (call it p') logically equivalent to p as follows. Let n = the number of Englishmen, and m = the number of children of Englishmen, then set

$$p' \equiv m/n = 2\frac{1}{2}.$$

The reductionist argues that accepting p as a true statement does not commit us to accepting the existence of that shadowy pseudo-entity, the average Englishman with his curious family, for p is reducible to a logically equivalent statement p′ which does not mention the average Englishman. Similarly, the reductionist would claim that statements apparently about numbers are reducible to logically equivalent statements which do not mention numbers. Thus we do not have to accept that numbers really exist, and these curious shadowy entities fortunately disappear. Frege does not consider this reductionist position, but, for the sake of completeness, we will discuss it briefly after we have given an account of Frege's own views.

First then let us consider Frege's crticism of the subjective view of number. Frege uses the term 'idea' to denote any particular content of an individual consciousness. In an article published late in his life, he explains his use of 'idea' as follows, (1918) Thoughts pp. 13-14:

> "Even an unphilosophical man soon finds it necessary to recognize an inner world distinct from the outer world, a world of sense-impressions, of creations of his imagination, of sensations, of feelings and moods, a world of inclinations, wishes and decisions. For brevity's sake I want to use the word 'idea' to cover all these occurrences, except decisions."

In this terminology, the subjective view of number can be formulated as the claim that numbers are ideas. The first of Frege's arguments against this position is that if numbers were indeed ideas, there would be not just one number two, but as many number twos as there were people who had learnt arithmetic, for each such person would have a different idea of two in his particular consciousness. This consequence of the subjective position, Frege regards as absurd and untenable. As he says, (1884) Foundations of Arithmetic § 27 p. 37e:

> "If the number two were an idea, then it would have straight away to be private to me only. Another man's idea is, *ex vi termini* (from the power of the boundary-line – D.G.), another idea. We should then have it might be many millions of twos on our hands. We should have to speak of my two and your two, of one two and all twos. If we accept latent or unconscious ideas, we should have unconscious twos among them, which would then return subsequently to consciousness. As new generations of children grew up, new generations of twos would continually be being born, and in the course of millennia these might evolve, for all we could tell, to such a pitch that two of them would make five."

Another argument of Frege's is that the subjectivists make arithmetic depend on psychology, whereas, surely arithmetic is more certain and exact than psychology. As he puts it, (1884) Foundations of Artihmetic, § 27, p. 38e:

> "It would be strange if the most exact of all the sciences had to seek support psychology, which is still feeling its way none too surely."

These arguments against the subjective view of number seem to me strong, if not decisive, and yet just this view was adopted by Brouwer in 1907 (23 years after Frege's critique) when laying the foundations of intuitionism. It is interesting then to ask whether Brouwer was able to counter Frege's criticisms, but it turns out that Brouwer never in fact mentions Frege. As Heyting says, (1975) Footnote in A. Heyting (ed.) L. E. J. Brouwer, Collected Works, Vol. 1, p. 568:

"Brouwer seems not to have known Frege's work. He has never mentioned it."

Let us now turn to Frege's own positive views. He holds that numbers do exist, though not as subjective, psychological ideas, but rather as objective entities. On the other hand, they are not physical objects in space and time. As he says, (1884) Foundations of Arithmetic, § 27, p. 38e:

"And we are driven to the conclusion that number is neither spatial and physical, like MILL's piles of pebbles and gingersnaps, nor yet subjective like ideas, but non-sensible and objective."

In another place Frege tries to explain further his notion of something which is objective without being tangible, (1884) Foundations of Arithmetic, § 26, p. 35e:

"I distinguish what I call objective from what is handleable or spatial or actual. The axis of the earth is objective, so is the centre of mass of the solar system, but I should not call them actual in the way the earth itself is so."

Frege then seems to recognize three sorts of things: (1) material objects in the external world, (2) ideas in a particular consciousness, and (3) objective abstract entities like numbers. Just as the natural sciences examine objects of type (1), so mathematics investigates objects of type (3). Indeed Frege draws a parallel between the two kinds of investigation, (1884) Foundations of Arithmetic, § 96, pp. 107e-8e:

"... even the mathematician cannot create things at will, any more than the geographer can; he too can only disover what is there and give it a name."

In his later article on 'Thoughts', Frege develops further this theory of a world or realm of abstract entities. As the title indicates, he is here concerned not with numbers, but with what he calls 'thoughts'. He takes a thought to be the content which is expressed by a sentence, and, as in the case of numbers, argues that thoughts cannot be (subjective) ideas, because, (1918) Thoughts, p. 16:

"... other people can assent to the thought I express in the Pythagorean theorem just as I do..."

Frege draws the following conclusion (op. cit. pp. 17-18):

"So the result seems to be: thoughts are neither things in the external world nor ideas.

A third realm must be recognized. Anything belonging to this realm has it in common with ideas that it cannot be perceived by the senses, but has it in common with things that it does not need an owner so as to belong to the contents of his

consciousness. Thus for example the thought we have expressed in the Pythagorean theorem is timelessly true, true independently of whether anyone takes it to
be true. It needs no owner. It is not true only from the time when it is dicovered;
just as a planet, even before anyone saw it, was in interaction with other planets."

Here Frege only explicitly mentions thoughts, but it seems reasonable to presume that numbers too would be inhabitants of his third
realm. This realm has obvious similarities to Plato's world of forms,
and it may strike many as being unpleasantly mysterious and
metaphysical. Is it not possible to dispense with this realm using the
reductionist strategy? This is what we must next consider.

Our hypothetical reductionist is prepared to admit material
objects in space and time, including plants, animals and human
beings. He is also perhaps prepared to admit subjective ideas in an
individual consciousness; but he is very reluctant to allow the existence of abstract entities such as numbers. In the language of modern
logic, he therefore holds that in the quantifiers $(\forall x)$ (for all x)
and $(\exists x)$ (there is an x), x should range only over material objects (or
possibly psychological objects such as thoughts), but not over abstract objects. The question is now whether we can develop ordinary
arithmetic and theory of numbers if we accept this prescription.

Certainly some numerical statements can still be made. Perhaps
the easiest way to do so is to introduce the numerical quantifiers
$(\exists_0 x)$, $(\exists_1 x)$, $(\exists_2 x)$, ... , $(\exists_n x)$, ... where $(\exists_n x)P(x)$ is supposed to mean:
'there are exactly n x such that $P(x)$'. These quantifiers can be de-
fined as follows:

$(\exists_0 x)P(x) =_{\text{def}} \neg (\exists x)P(x)$

$(\exists_1 x)P(x) =_{\text{def}} (\exists x)(P(x) \wedge (\forall y)(P(y) \rightarrow y = x))$

$(\exists_2 x)P(x) =_{\text{def}} (\exists x)(\exists y)(P(x) \wedge P(y) \wedge x \neq y$
$\wedge (\forall z)(P(z) \rightarrow (z = x \vee z = y)))$

. . .

They enable us to make numerical statements such as 'There are five
apples on that table' which becomes simply: $(\exists_5 x)P(x)$ where $P(x) =_{\text{def}} x$ is an apple on that table.

We can also use the numerical quantifiers to express simple
arithmetical equations such as $5 + 7 = 12$. In fact this becomes:

$((\exists_5 x)P(x) \vee (\exists_7 y)Q(y) \wedge \neg (\exists z)(P(z) \wedge Q(z)))$
$\rightarrow (\exists_{12} w)(P(w) \vee Q(w))$

However, as soon as we go on to even quite simple statements of
the theory of numbers, the picture changes. Consider, for example,
the statement that there is a prime number greater than any given

number. If we write: '*m* is a prime number', as Pr(*m*), this becomes

$$(\forall n)(\exists m)(\mathrm{Pr}(m) \wedge m > n) - (*)$$

However this involves quantification over numbers which, since they are abstract entities, our reductionist cannot allow. Moreover there seems to be no way, using some device such as numerical quantifiers, of reducing (*) to a form in which there is only quantification over material objects. The same applies to the principle of mathematical induction:

$$(P(0) \wedge (\forall n) (P(n) \to P(n+1)))) \to (\forall n)P(n)$$

Here again we have a quantification over numbers which cannot be eliminated.

Thus if we take our reductionist's prescriptions seriously, we would be forced to abandon virtually all higher number theory, and most of Peano arithmetic. We would not be able to get much beyond simple numerical equations. So, if we are going to accept at least a reasonable portion of modern mathematics, reductionism will not work, and we will have to accept a world of abstract entities such as numbers whether we like it or not. Let us therefore examine this "third realm" a little more closely.

Recently Popper has followed Frege in distinguishing three worlds. The first world is of material objects in space and time; the second world is of psychological entities — ideas in Frege's sense; and the third world is of abstract entities. However, Popper's account of the third world differs in an interesting and significant fashion from Frege's and Plato's. As Popper himself says, (1972) Objective Knowledge, Ch. 3, p. 122:

"Plato's third world was divine; it was unchanging and, of course, true. Thus there is a big gap between his and my third world; my third world is man-made and changing. It contains not only true theories but also false ones, and especially open problems, conjectures, and refutations."

So Popper holds that the world of abstract entities is man-made and changes in time, while both Frege and Plato hold that this world exists in a timeless sense independently of human beings. Frege indeed explicitly criticizes the view that numbers might change with time. He writes, (1884) Foundations of Arithmetic, Introduction, p. VIe:

"... astronomers would hestitate to draw any conclusions about the distant past, for fear of being charged with anachronism, — with reckoning twice two as four regardless of the fact that our idea of number is a product of evolution and has a history behind it, It might be doubted whether by that time it had progressed so far. How could they profess to know that the proposition $2 \times 2 = 4$ was already in existence in that remote epoch?"

Frege's argument here is fallacious. Astronomers apply our present day concept of number when describing the distant past — but that does not imply either that the concept of number existed then or that it is timeless. In the same way some astronomers use the English language to describe the distant past without implying that English existed then or that it is timeless.

Of the views so far considered regarding the existence of numbers and other abstract entities, it seems to me that Popper's is the most plausible. I propose to call his view *constructive Platonism,* because it is, in a sense, a synthesis of the views of Brouwer on the one hand and of Frege and Plato on the other. Brouwer held that numbers are the subjective mental constructions of particular individuals. The claim that numbers are subjective and mental has, in my view, been decisively refuted by Frege. We are thus led to accept Platonism in so far as it claims that numbers are objective but yet not material. On the other hand the traditional Platonic view of numbers as timeless and existing independently of human beings seems implausible. Brouwer does seem to be right in holding that numbers are human constructions. They are not however, subjective mental constructions, but rather objective social constructions. Just as human beings in the course of their history gradually learnt how to construct material artefacts such as tools and houses, so they learnt how to construct non-material artefacts such as numbers. Indeed these non-material artefacts (numbers) can be used in conjunction with material artefacts (tools) to construct further material artefacts (houses).

Although constructive Platonism seems to me the most plausible theory of abstract entities, it is useless to pretend that it does not contain problems and difficulties. To begin with abstract entities such as numbers still appear as strange and mysterious, and more clarification regarding their nature is needed. Moreover we have to explain how exactly human beings do construct abstract entities. Houses are built by adding brick to brick; but how is the number three constructed?

In addition to these general points, there is a specific difficulty recently pointed out by Currie who writes, (1978) The Objectivism of Frege and Popper, Ch. 5, pp. 244-5:

> "If a choice is to be made about what is the correct set theoretic structure, it cannot be made by claiming that there is a unique abstract structure in the third world which makes our axioms true or false. Popper's third world is pluralistic in a sense in which the traditional platonic heaven is not. It is teeming with alternatives: standard and non-standard models of our theories, intuitionistic arithmetic as well as its classical rival, logics with all conceivable values, etc."

The point here is that, on the traditional Platonic view, we can regard a mathematical proposition as true if it corresponds to what is actually the case in the timeless world of forms. But if this world of forms is a human creation, it must, as Popper explicitly states, contain false theories as well as true ones, and misleading and erroneous conceptions as well as correct ones. Thus mathematical truth can no longer be analysed as correspondence with what holds in the Platonic world of abstract entities, and some other account is needed.

I believe that the problems involved in constructive Platonism can be overcome, but to investigate this matter further would take us too far from our present concerns. In the next chapter therefore we will return to our exposition of Frege's views.

Chapter 7.
Frege's Logicism

We have already remarked that, in order to establish his logicist thesis, Frege has to give a definition of number in terms of purely logical notions, and to show that all arithmetical reasoning, particularly reasoning from n to $n+1$, or the principle of mathematical induction, can be reduced to logical inference. Let us now examine more closely how he sets about these tasks.

Frege begins his attempt to define number in (1884) Foundations of Arithmetic § 55 p. 67ᵉ. He first claims to have established that the content of a statement of number is an assertion about a concept, and therefore goes on to define: '0 belongs to F', 1 belongs to F', ..., 'n belongs to F', ... where F is an arbitrary concept. His definitions, translated into the notation of modern mathematical logic, are as follows (op. cit. § 55 p. 56ᵉ):

'0 belongs to F' $=$ def $(\forall a)\neg F(a)$

'1 belongs to F' $=$ def $(\exists a)F(a) \wedge (\forall a)(\forall b)(F(a) \wedge F(b) \to a = b)$

. . .

'$n+1$ belongs to F' $=$ def $(\exists a)(F(a) \wedge n$ belongs to the concept '$F(x) \wedge x \neq a$')

It will be seen that these definitions are essentially the same as those of the numerical quantifiers $(\exists x)$, $(\exists x)$, ..., $(\exists x)$, ... which we gave in chapter 6.

Do these formulae constitute an adequate definitions of number? Frege answers 'no', for the following reasons (op. cit. § 56 p. 68ᵉ):

"... we can never − to take a crude example − decide by means of our definitions whether any concept has the number JULIUS CAESAR belonging to it, or whether that same familiar conqueror of Gaul is a number or not. Moreover we cannot by the aid of our suggested definitions prove that, if the number a belongs to the concept F and the number b belongs to the same concept, then necessarily $a = b$."

Our earlier discussion of the numerical quantifiers should, I hope, help to clarify Frege's point here. We considered a reductionist programme for eliminating abstract entities, and expressing arithmetical propositions by the numerical quantifiers $(\exists x)$ where x ranges only over material (and possibly psychological) entities. We argued that such a reductionist programme does not work, and that to obtain the usual mathematical theory of numbers we need to quantify over numbers, and treat numbers as objects. This is essentially what Frege is saying here. His definitions of '0 belongs to F', '1 belongs to F', ..., '*n* belongs to F', ... do not allow us to treat numbers as objects, and so are inadequate for arithmetic.

But how then we introduce numbers as objects, since we cannot exhibit numbers in sense-perception (or Kantian intuition)? Frege states that we can do so by laying down identity criteria for numbers i.e. by defining what is meant by saying that two numbers are equal, (1884) Foundations of Arithmetic § 62 p. 72^e:

> "In our present case, we have to define the sense of the proposition
> "the number which belongs to the concept F is the same as that which belongs to the concept G";..."

To illustrate his procedure, Frege considers the problem of defining: 'the direction of the line a'. We begin by defining:

the direction of line a = the direction of line b if and only if $_{def}$ line a is parallel to line b (a ∥ b)

The notion of parallel lines, Frege thinks, must be given originally in intuition like "everything geometrical" (op. cit. § 64 p.75^e). However to show that the definition is legitimate, we have to show that it satisfies the analytic truths about the notion of identity — which truths Frege takes to be summed up in Leibniz' Law i.e. *eadem sunt, quorum unum potest substitui alteri salva veritate*.[1] This amounts to showing that ∥ is an equivalence relation.

Having defined 'the direction of line a = the direction of line b', we now have to define 'the direction of line a' (*tout court*) in order to be able to decide whether e.g. England is the direction of the earth's axis. Frege suggests the following (op. cit. § 68 p. 79^e):

> "the direction of line a is the extension of the concept "parallel to line a";"

(i.e. is the set of all lines parallel to a)

To define 'the number which belongs to the concept F' we proceed in the same way. We say that two concepts F and G are equal if the things that fall under them can be put in 1-1 correspondence. Then we define (op. cit. § 68, pp. 79^e-80^e):

> "the Number which belongs to the concept F is the extension of the concept "equal to the concept F"."

So a number is a set of concepts.

Frege admits that it is somewhat curious to identify a number with the extension of a concept, and this may lead to strange ways of

---

[1] They are the same, one of which can be substituted for the other preserving truth.

speaking. However nothing incorrect will result, and he doesn't regard such objections as very serious.

To show that the notion of 1-1 correspondence does not presuppose that of number, Frege gives the following example (op. cit. § 70 pp. 81e-82e):

"If a waiter wishes to be certain of laying exactly as many knives on a table as plates, he has no need to count either of them; all he has to do is to lay immediately to the right of every plate a knife, taking care that every knife on the table lies immediately to the right of a plate."

Frege next shows that 1-1 correspondence can be defined in terms of purely logical notions, and that 1-1 correspondence is an equivalence relation. This is pretty familiar material nowadays.

Frege next defines (op. cit. § 72 p. 85e):

"the expression
"*n* is a Number"
is to mean the same as the expression
"there exists a concept such that *n* is the Number which belongs to it"."

He then goes on to define the individual numbers 0, 1, 2, ..., and the relations of successor. This he does as follows:

Definition of 0 (op. cit. § 74 p. 87e)

"Since nothing falls under the concept "not identical with itself", I define nought as follows:
0 is the Number which belongs to the concept "not identical with itself"."

Definition of Successor (op. cit. § 76 p. 89e)

"I now propose to define the relation in which every two adjacent members of the series of natural numbers stand to each other. The proposition:
"there exists a concept F, and an object falling under it *x*, such that the Number which belongs to the concept F is *n* and the Number which belongs to the concept 'falling under F but not identical with *x*' is *m*"
is to mean the same as
"*n* follows in the series of natural numbers directly after *m*"."

Definition of 1 (op. cit. § 77 p. 90e):

"1 is the Number which belongs to the concept "identical with 0","

It can then be proved that 1 follows in the series of natural numbers directly after 0.

Frege's next task is to prove a number of propositions regarding the relation just defined viz. "n follows in the series of natural numbers directly after m" or, briefly, m is a successor of n. Here are three examples:

(1) '*m* is a successor of *n*' is a 1-1 relation.

(2) Every number except 0 is a successor of a number.

(3) Every number has a successor.

In order to prove (3), Frege has, given a number *n*, to find a concept whose extension has $n+1$ members. In fact he chooses the concept: "member of the series of natural numbers ending with n". But to define this concept, Frege has to define the notions of "series" and "following in a series". He remarks that in his *Begriffsschrift* he has in fact defined: "y follows x in the φ-series" in purely logical terms, and proved some properties of it. The definition amounts to saying that the relation holds if y has all the φ-hereditary properties possessed by the immediate successors of x. He goes on to say (op. cit. § 80 p. 93ᵉ):

> "Only by means of this definition of following in a series is it possible to reduce the argument from n to $(n+1)$, which on the face of it is peculiar to mathematics, to the general laws of logic."

Frege does not show how this reduction is accomplished, but he does give a sketch of a proof that every number has a successor.

What conclusion does Frege draw from all this. He writes as follows, (1884) Foundations of Arithmetic § , p. 99ᵉ:

> "I hope I may claim in the present work to have made it probable that the laws of arithmetic are analytic judgements and consequently a priori. Arithmetic thus becomes simply a development of logic, and every proposition of arithmetic a law of logic, albeit a derivative one. To apply arithmetic in the physical sciences is to bring logic to bear on observed facts; calculation becomes deduction. The laws of number will not, as BAUMANN thinks, need to stand up to practical tests if they are to be applicable to the external world;..."

It is important to notice that Frege only claims "to have made it probable that the laws of arithmetic are analytic judgements". In effect he has sketched how to define number in terms of purely logical notions (i.e. for him 'concept', 'identity', $(\forall x)$, etc.), and how the theorems of arithmetic can be deduced using only logical principles and without appeal to any special mathematical inferences. However, as he goes on to say, some doubts might remain as to whether there might be a gap in the chain of proofs — a gap which could only be filled by appeal to some non-logical principle. (op. cit. § 90 p. 102ᵉ):

> "I do not claim to have made the analytic character of arithmetical propositions more than probable, because it can still always be doubted whether they are deducible solely from purely logical laws, or whther some other type of premiss is not involved at some point in their proof without our noticing it."

In effect Frege has set himself a programme: to write out in full his definitions and proofs, and thus to establish, beyond doubt, the analytic character of arithmetic. It was to carry out this programme that he invented his *Begriffsschrift* (Concept Writing) because ordinary language was not sufficiently precise for the purpose. As he says (op. cit. § 91, p. 103ᵉ):

"To minimize these drawbacks, I invented my *Begriffsschrift*. It is designed to produce expressions which are shorter and easier to take in, and to be operated like a calculus by means of a small number of standard moves so that no step is permitted which does not conform to the rules which are laid down once and for all. It is impossible, therefore, for any premiss to creep into a proof without being noticed."

Frege spent the next nineteen years carrying out this long and difficult programme. The results were: *Grundgesetze der Arithmetik, begriffsschriftlich abgeleitet* (Fundamental Laws of Arithmetic, derived using the concept writing), Vol. I. 1893. Vol. II. 1903. We shall examine in chapter 12 the fate which overtook this work. In the next two chapters we shall consider Dedekind's version of logicism, and see how it compares with Frege's.

Chapter 8.
Dedekind and Set Theory

Dedekind's main work on the foundations of arithmetic: *Was sind und was sollen die Zahlen?* (literally: what are and what ought to be the numbers?) was published in 1888, that is four years after Frege's Foundations of Arithmetic. However Dedekind worked out his ideas independently of Frege, and only came across Frege's book after his own was completed. In the second edition of '*Was sind und was sollen die Zahlen?*' (1893), Dedekind says in the preface (pp. 42-43):

> "About a year after the publication of my memoir I became acquainted with G. Frege's *Grundlagen der Arithmetik,* which had already appeared in the year 1884. However different the view of the essence of number adopted in that work is from my own, yet it contains, particularly from § 79 on, points of very close contact with my paper, especially with my definition (44). The agreement, to be sure, is not easy to discover on account of the different form of expression; but the positiveness with which the author speaks of the logical inference from n to $n+1$ (page 93, below) shows plainly that here he stands upon the same ground with me."

In a private letter of 1890 quoted by Hao Wang in his interesting article on The Axiomatization of Arithmetic, Dedekind warns his correspondent (op. cit. p. 151):

> "Only one must not be put off by his (i.e. Frege's) somewhat inconvenient terminology."

Later on we will examine the specific point of similarity between § 79 of Frege's *Grundlagen,* and (44) of Dedekind's *Was sind und was sollen die Zahlen?* Let us begin, however, with some rather more general issues. The main point of similarity with Frege is that Dedekind also espouses logicism (the view that arithmetic can be reduced to logic). This is how he puts it in the Preface to the first edition, (1888) *Was sind und was sollen die Zahlen?* pp. 31-32:

> "In speaking of arithmetic (algebra, analysis) as a part of logic I mean to imply that I consider the number-concept entirely independent of the notions or intuitions of space and time, that I consider it an immediate result from the laws of thought. My answer to the problem propounded in the title of this paper is, then, briefly this: numbers are free creations of the human mind; they serve as a means of apprehending more easily and more sharply the difference of things. It is only through the purely logical process of building up the science of numbers and by

thus acquiring the continuous number-domain that we are prepared accurately to investigate our notions of space and time by bringing them into relation with this number-domain created in our mind."

So Dedekind like Frege rejects the Kantian theory of arithmetic. However we note at once one difference from Frege. Dedekind has a psychologistic rather than Platonistic view of logic. He speaks of "the laws of thought" and of numbers as being "free creations of the human mind". Another difference between the two emerges later. Dedekind regards the notion of class, or set, or, to use his own terminology, system, as a logical notion. But Frege denies this. In his preface to (1893) *Grundgesetze der Arithmetik* Vol. 1. Preface. Furth Translation. p. 4, Frege writes:

> "Herr Dedekind, like myself, is of the opinion that the theory of numbers is a part of logic; but his work hardly contributes to its confirmation, because the expressions "system" and "a thing belongs to a thing", which he uses, are not usual in logic and are not reduced to acknowledged logical notions."

For Frege, "concept" and "extension of a concept" are logical notions, whereas "set", "class", "system" are not. Thus Frege's point of view leads to higher-order logic and type theory; whereas Dedekind's leads to axiomatic set theory.

Another difference is that Frege feels the need for formal logic (his *Begriffsschrift*) in order to write out the requisite proofs completely, whereas Dedekind proceeds informally. Frege himself makes this point, (1893) *Grundgesetze der Arithmetik* Vol I Preface Furth Translation p. 4:

> "... if we compare Herr Dedekind's work, *Was sind und was sollen die Zahlen?*, the most thoroughgoing work on the foundations of arithmetic that has lately come to my notice. In much less space it pursues the laws of arithmetic much further than is done here. To be sure, this brevity is attained only because a great deal is really not proved at all. ... an inventory of the logical or other laws taken by him as basic is nowhere to be found and even if it were, there would be no way of telling whether no others were actually used; for that to be possible the proofs would have to be not merely indicated but carried out, without gaps."

We see here, once again, Frege's obsessive desire to prove beyond doubt that arithmetic is reducible to logic by exibiting all the axioms and rules of inference needed, and thereby showing that they are all logical in character so that no appeal to intuition or empirical considerations is needed. Dedekind, though a logicist, has not the same over-ruling passion to demonstrate his position conclusively, and is content with the usual informal mathematical standards of rigour. As a result his work has much more mathematical elegance than Frege's.

Although Dedekind does not use formal logic, the claim that "an inventory of the logical or other laws taken by him as basic is nowhere to be found" is not altogether true, since, as we shall see, Dedekind does state some of the basic principles of set theory, which for him are part of logic.

Finally Frege has lengthy philosophical discussions of Kant, Mill, etc., whereas Dedekind sticks for the most part to mathematics. Generally speaking Dedekind is more of a mathematician than a philosopher.

Let us now begin a more detailed analysis of Dedekind's (1888) *Was sind und was sollen die Zahlen?* As already remarked, Dedekind takes the notion of *system* as basic, and he introduces it as follows (op. cit. (2) p. 45):

> "It very frequently happens that different things, *a, b, c,* ... for some reason can be considered from a common point of view, can be associated in the mind, and we say that they form a *system* S; ..."

The elements of psychologism in this passage produced a characteristic expostulation from Frege, (1893) *Grundgesetze der Arithmetik*, Furth translation, pp. 29-30:

> "For an undertaking of this kind to succeed, it is of course necessary that we grasp precisely the concepts required. This applies particularly to what the mathematicians would like to designate by the word "set". Dedekind uses the word "system" with very much the same purpose. But despite the explanations in my *Grundlagen* four years earlier, a clear insight into the essence of the matter is not to be found in Dedekind, although he occasionally comes near the mark, ... But other passages wander off again, for example the following (pp. 1-2): "It very frequently occurs that different things *a, b, c,* ..., regarded for some reason from a common point of view, are put together in the mind; and we say then that they form a *system* S." Here a presentiment of the truth is indeed contained in the 'common point of view'; but this 'regarding', this 'putting together in the mind', is not an objective characteristic. I ask, in whose mind? If they are put together in one mind but not in another, do they form a system then? What is supposed to be put together in my mind, no doubt must be in my mind: then do the things outside myself not form systems? Is a system a subjective figure in the individual soul? In that case is the constellation Orion a system? And what are its elements? The stars, or the molecules, or the atoms?"

Here we find again Frege's characteristic arguments against psychologism, but it should be observed that there is nothing inherently psychologistic about the notion of 'set' or 'system'. To take Frege's own example, the set of stars in Orion is something which exists in the objective, material world quite independently of human consciousness.

Having introduced the basic notion of system, Dedekind proceeds in the course of his exposition to state various principles which this notion satisfies. These principles are not explicitly introduced as axioms, but they nonetheless bear a close relation to the later axioms of set theory. Indeed, as we shall see, Dedekind's (1888) *Was sind und was sollen die Zahlen?* is the principal source for Zermelo's (1908) Investigations in the Foundations of Set Theory I, and in fact Zemelo frequently refers to Dedekind in this paper.

Zemelo takes the notion of set as an undefined notion. He remarks laconically, (1908), Foundations of Set Theory I § 1,1 p. 201:

"Set theory is concerned with a *domain* **B** of individuals, which we shall call simply *objects* and among which are the *sets*."

Thus Dedekind's psychologism is avoided.

Zemelo's Axiom I (Axiom of Extensionality) appears in Dedekind (1888) *Was sind und was sollen die Zahlen?* (2) p. 45 as follows:

"The system S is hence the same as the system T, in symbols S = T, when every element of S is also element of T, and every element of T is also element of S."

Dedekind remarks in a footnote that, contrary to Kronecker, whether *s* is a member of S need not be determined effectively.

Dedekind next introduces the notion of 'part' as follows (op. cit. (3) p. 46):

"Definition. A system A is said to be *part* of a system S when every element of A is also element of S."

This Dedekind writes A ∋ S, which corresponds to the moderm A ⊆ S except for one thing. Nowadays it is standard to distinguish between a is a member of S (*a* ε S), and A is subset of S (A ⊆ S). However Dedekind conflates these two notions. He writes (op. cit. (3) p. 46):

"Since further every element *s* of a system S by (2) can be itself regarded as a system, we can hereafter employ the notation *s* ∋ S."

This amounts to identifying an element *s* with its unit set {*s*}.

Peano was the first to distinguish between *a* is a member of *b* (in his notation *a* ε *b*), and *a* is contained in *b* (in his notation *a* ⊃ *b*). The notations ε, ⊃ are given in his (1889) Arithmetices Principia. However, despite having a difference in notation, Peano is not yet clear about the conceptual difference. Indeed the formula 56 of the section entitled 'Notations of Logic' (op. cit. p. 108) states the following: if *s* is a class and *k* is a subset of *s* (*k* ⊃ *s*), then *k* is a member of *s* (*k* ε *s*) iff *k* has one and only one member. Later however, in his (1894) *Notations de Logique Mathématique,* § 31, p. 160, Peano was to distinguish between an object *x*, and its unit class (in his notation ιx), and thus to clear the whole matter up.

Dedekind excludes the empty set. He writes (1888) *Was sind und was sollen die Zahlen?* (2) pp. 45-6:

"... we intend here for certain reasons wholly to exclude the empty system which contains no element at all, although for other investigations it may be appropriate to imagine such a system."

Dedekind does not explain what "certain reasons" he has in mind.

However the whole matter may be quite closely connected with Dedekind's failure to distinguish between $a \, \varepsilon \, S$ and $A \subseteq S$. Considerable light is shed on this question by Frege's critical review of Schröder's *Algebra der Logik*. Schröder, like Dedekind, fails to distinguish between ε and $\subseteq$, but, unlike Dedekind, he does introduce the empty set. Some difficulties consequently appear which Frege seeks to resolve.

Frege distinguishes what he calls a *domain-calculus* from a more general interpretation of class. In a domain-calculus, the domains are considered as physical objects, and we have only the part-whole relation. If we take class in the sense of domain, there is no empty class. As Frege himself puts it, (1895) A Critical Elucidation of Some Points in E. Schröder's *Vorlesungen über die Algebra der Logik* p. 89:

"A class, in the sense in which we have so far used the word, consists of objects; it is an aggregate, a collective unity, of them; if so, it must vanish when these objects vanish. If we burn down all the trees of a wood, we thereby burn down the wood. Thus there can be no empty class."

In a more general treatment of classes, however, we must distinghuish A is a subset of S (which Frege writes A sub S) from a is a member of S (which he writes a subter S). Frege remarks in a footnote (op. cit. p. 94) that his 'sub' and 'subter' correspond to Peano's '$\supset$' and 'ε'. The distinction between the two situations is made as follows (op. cit. pp. 92-3):

"We see again from this that we are no longer standing on the basis of the domain-calculus; for there we had only the part-whole relation, and there was no ground for this distinction between the cases where a class contains something as an individual and where it contains it as a class."

Now suppose we have a domain-calculus, and, like Schröder but unlike Dedekind, we allow the empty set $\emptyset$. Difficulties do arise, as Schröder himself realized. I shall next explain these using modern notation. We have of course $\emptyset \subseteq S$, for any set S. So setting $S = \{a_1, a_2, \ldots, a_n\}$, we have $\emptyset \subseteq \{a_1, a_2, \ldots, a_n\}$. But now if we don't distinguish between $\subseteq$ and ε, it seems that we must have $\emptyset \, \varepsilon \, \{a_1, a_2, \ldots, a_n\}$. Thus $\emptyset$ always "slips in" to any set. But now we can ask: 'how many members does $\{a_1, a_2, \ldots, a_n\}$ have?' Intuitively it would seem that it should have n members, but, if $\emptyset$ slips in as well, perhaps we should say that it should have $n + 1$ members.

It seems to me likely that Dedekind was aware of difficulties of this kind, and that they constituted his "certain reasons" for excluding the empty set. However, as Frege goes on to show, the problem is resolved by distinguishing between 'ε' and '$\subseteq$'. $\emptyset \subseteq S$ for any set S, but it is not true that $\emptyset \, \varepsilon \, S$ for any set S. As we have already seen,

Frege uses Schröder's difficulties here as another argument for taking 'concept' as more basic than 'set'. But his own successful resolution of the problem seems to me to show that this last stage of the argument is *non-sequitur*. Once the distinction between 'ε and '⊆' had been clearly made (and this was no easy task historically), then the difficulties concerned with the empty set disappeared.

As the first part of his Axiom II (Axiom of elementary sets), Zermelo gives, (1908) Foundations of Set Theory I, § 1, 4, p. 202:

"There exists a (fictitious) set, the *null set*, 0, that contains no element at all."

Thus Zermelo, unlike Dedekind, explicitly postulates the empty set. On the other hand, there is a trace of Dedekind's influence in that Zermelo regards the empty set as "a (fictitious) set". Really, from the point of view of his set theory, the empty set is no more, or less, fictitious than any other set.

In (1888) *Was sind und was sollen die Zahlen?* I (8) pp. 46-47, Dedekind gives a definition of the union of any arbitrary set of sets A, B, C, ... This definition corresponds to Zermelo's Axiom V (Axiom of the Union) (Zermelo (1908) Foundations of Set Theory § 1, 10, p. 203). Dedekind also gives a definition of the intersection of any set of sets A, B, C, ... (op. cit. I (17) pp. 48-49). Zermelo does not need to introduce a special axiom for this, since the existence of the intersection follows from his other axioms. Dedekind, unlike Zermelo, has to stipulate that the sets A, B, C, ... should have at least one common element, before he can apply his definition of intersection.

Let us now turn to Dedekind's treatment of infinite sets, which is one of the most interesting parts of his monograph. In (1888) *Was sind und was sollen die Zahlen?* V (64) p. 63, Dedekind gives his famous definition of infinite set:

"Definition. A system S is said to be *infinite* when it is similar to a proper part of itself ...; in the contrary case S is said to be a *finite* system."

('Similar' is here used in the sense that two sets are similar if they can be put in 1-1 correspondence).

Dedekind now goes on to give the following rather curious proof of the existence of at least one infinite set (op. cit. V (66) p. 64):

"Theorem. There exist infinite systems.

Proof. My own realm of thoughts, i.e., the totality S of all things, which can be objects of my thought, is infinite. For if s signifies an element of S, then is the thought s', that s can be object of my thought, itself an element of S."

The map $s \rightarrow s'$ is now a similarity mapping (i.e. a 1-1 correspondence) of S onto a proper subset of itself, and so S is infinite. Dedekind mentions in a footnote that (op. cit. V (66) p. 64):

"A similar consideration is found in § 13 of the *Paradoxien des Unendlichen* by Bolzano (Leipzig, 1851)."

This argument is not very convincing — particularly in the light of the set-theoretic paradoxes which were discovered later. However it is interesting to try to see what is wrong with it. When Dedekind speaks of his realm of thoughts, he clearly means the set of his possible thoughts, since the set of his actual thoughts was presumably finite (at least during his earthly existence). However the notion of 'possible thought' is somewhat obscure, and it is not clear that there are such things. This is an objection which Russell makes in his discussion of Dedekind's argument (1919) Introduction to Mathematical Philosophy Ch. XIII. The Axiom of Infinity and Logical Types. pp. 138-140). Russell writes (op. cit. p. 139):

> "If the argument is to be upheld, the "ideas" intended must be Platonic ideas laid up in heaven, for certainly they are not on earth. But then it at once becomes doubtful whether there are such ideas."

But even if we allow "possible thoughts", there is another difficulty. We can presumably have a possible thought of any possible thing. Thus the cardinal number of the set of possible thoughts must be greater than or equal to the cardinal number of the set of possible things. But now we are into Cantor's paradox. Let V = the set of possible things. Then V should presumably have the greatest cardinal number. But if PV = the power set of V = the set of all subsets of V, then by Cantor's theorem the cardinal number of PV is strictly greater than the cardinal number of V.

Zermelo rejects Dedekind's proof of the existence of an infinite set for essentially the reason just given. He writes, (1908) Foundations of Set Theory § 1, 13 p. 204 Ftnte 8:

> "The "proof" that Dedekind there attempts to give of this principle cannot be satisfactory, since it takes its departure from "the set of everything thinkable", whereas from our point of view the domain **B** itself, according to No. 10, does *not* form a set."

Accordingly Zermelo postulates the existence of an infinite set as his Axiom VII (Axiom of Infinity). Nonetheless he says (op. cit. p. 204) that this axiom "is essentially due to Dedekind". His meaning may be that Dedekind's failure to give an *a priori* proof of the existence of an infinite set showed the need for an axiom postulating the existence of such a set.

Let us call a set which is infinite in Dedekind's sense, *Dedekind-infinite*. Once we have introduced the natural numbers 1, 2, 3, ... ,n, ... and the sets $Z_n = \{1, 2, 3, ... ,n\}$, we can define a set S to be *ordinary-infinite* if for all n, Z_n can be mapped 1-1 into S. We then have the theorem that a set S is Dedekind-infinite if and only if it is ordinary-infinite. This is in fact theorem 159 of *Was sind und was sollen die Zahlen?* (XIV p. 105).

The proof that Dedekind-infinite → ordinary-infinite, is elementary, but the converse in fact needs the Axiom of Choice which Dedekind assumes implicitly.

Dedekind does however remark (op. cit. XIV (159) p. 106):

"The proof of the converse – however obvious it may appear – is more complicated."

and in the very next sentence he makes a tacit appeal to the axiom of choice:

"If every system Z_n is similarly transformable in Σ, then to every number n corresponds such a similar transformation α_n of Z_n that $\alpha_n (Z_n) \ni \Sigma$."

For each n, Dedekind selects a particular similar transformation α_n, and this requires the axiom of choice.

The axiom of choice was first explicitly formulated by Zermelo in his 1904 paper: Proof that every set can be well-ordered – though it is mentioned in passing by Peano in 1890 and Beppo Levi in 1902[1]. In Zermelo's 1908 paper It appears as Axiom VI.

Zermelo's 1908 paper contains seven axioms for set theory. So far we have considered the following: Axiom I (Axiom of Extensionality), Axiom II (First part – postulation of existence of the empty set), Axiom V (Axiom of the union), Axiom VI (Axiom of Choice), and Axiom VII (Axiom of Infinity). In each case we have shown some connection between the axiom and Dedekind's 1888 monograph. For the sake of completeness, let us briefly consider Zermelo's remaining axioms i.e. Axiom II (second and third parts), Axiom III, and Axiom IV.

Axiom IV is the axiom of the power set. In states that to every set A, there corresponds the power set of A (PA) i.e. the set of all subsets of A. The principal use of this axiom is in the development of Cantor's theory of infinite cardinal numbers. Since Dedekind does not consider infinite cardinal numbers in his monograph, he has no need to introduce the concept of power set. Zermelo's Axioms II and III are really designed to replace the intuitive so-called Axiom of Comprehension. This states that given any property there exists the set of all things which have that property, or, in symbols:

$$(\exists y)(\forall x)(x \,\varepsilon\, y \leftrightarrow P(x)\ )-(*)$$

[1] For details see A. A. Fraenkel and Y. Bar-Hillel (1958) Foundations of Set Theory. Ch. II. § 4 pp. 47-48.

As we shall see in Ch. 12, this axiom leads at once to Russell's paradox. Zermelo's Axiom II (Axiom of Elementary Subsets) which postulates the empty set, the unit set $\{a\}$ of any object a, and the unordered pair set $\{a,b\}$ of any objects a,b, and his Axiom III (Axiom of Separation) are weaker substitutes for the axiom of comprehension designed to allow the development of set theory while avoiding contradiction. It is interesting in this context to observe that Dedekind does make an explicit appeal to the erroneous principle of comprehension. His proof of theorem 60 ((1888) *Was sind und was sollen die Zahlen?* IV (60) p. 62) begins:

"Indeed, if we denote by Σ the system of all things possessing the property E ..."

Since the property E is here arbitrary, Dedekind assumes that given any property E there is a set of all things possessing E i.e. the axiom of comprehension. This axiom, incidentally, not only leads to Russell's paradox, but establishes the existence of the empty set — contrary to Dedekind's earlier assertion. For setting $x \neq x$ for P(x) in (∗), we obtain the set of all x's such that $x \neq x$ ($\hat{x}(x \neq x)$) i.e. the empty set.

So far then we have described the set-theoretic framework of Dedekind's monograph, and shown that this bears a remarkably close relationship to later axiomatic set theory. We must next examine how Dedekind develops arithmetic within this framework.

Chapter 9.
Dedekind's Development of Arithmetic

It will be recalled that Dedekind endorsed Frege's contention that the inference from n to $n + 1$ is logical in character. What this claim means can perhaps be most easily seen within Dedekind's framework. In fact it amounts to this: the principle of mathematical (or complete) induction is a provable theorem of set theory once natural numbers have been defined in terms of sets.

Dedekind does not, however, proceed directly to a definition of natural number. His method is first to produce what might be considered as a set-theoretic generalisation of the sequence of natural numbers. He then proves, as a theorem, a corresponding generalisation of the principle of mathematical induction. The ordinary principle of mathematical induction then follows as a special case once natural numbers have been defined.

To see the motivation for Dedekind's definitions, let us consider the set N of natural numbers $= \{1, 2, 3, ... ,n, ...\}$ where, following Dedekind, we will take the sequence as beginning with 1. With each number n, we can associate its successor $n' = n + 1$. To generalize, let us take any arbitrary set S. Instead of the successor operation $n \rightarrow n'$, we can take any abritrary 1-1 transformation (ϕ say) of S into itself. If $a \, \varepsilon \, S$, we write $\phi(a) = a'$, and if $A \subseteq S$, $\phi(A) = A'$. Now in the case of natural numbers, we have $N' \subseteq N$. Let us therefore define a chain K ($=$ Kette) to be a subset of S such that $K' \subseteq K$. (Dedekind (1888) *Was sind und was sollen die Zahlen?* IV Definition (37) p. 56). However N is (intuitively) the smallest chain which contains the number 1. Analogously therefore we define the *chain of A* (written A_0), where A is any subset of S, to be the intersection of all chains containing A. (Dedekind op. cit. IV Definition (44) pp. 57-58).

We have already quoted a passage (Dedekind op. cit. p. 42), in which Dedekind points out an analogy between his definition (44) and § 79 of Frege's *Grundlagen der Arithmetik*. In that section, Frege gives his definition of "y follows in the ϕ-series after x". Correspondingly Dedekind's A_0 can be considered as "the φ-series beginning with A". Both Frege and Dedekind are trying to abstract from the sequence of natural numbers to produce a more general notion of "φ-sequence".

Dedekind uses his notion of "the chain of A" (A_0) to prove his generalisation of the principle of mathematical induction. This is in fact the following theorem, Dedekind (1888) *Was sind und was sollen die Zahlen?* IV (59) p. 60:

> "59. Theorem of complete induction. In order to show that the chain A_0 is part of any system Σ — be this latter part of S or not — it is sufficient to show,
>
> ρ. that $A \ni \Sigma$, and
>
> σ. that the transformation of every common element of A_0 and Σ is likewise element of Σ."

He then remarks (op. cit. IV (60) p. 61):

> "The preceding theorem, as will be shown later, forms the scientific basis for the form of demonstration known by the name of complete induction (the inference from n to $n + 1$); ..."

Dedekind uses theorem 59 to justify both proofs by induction, and definitions by induction.

Frege's reduction of mathematical induction to logical inference is similar, except, for Dedekind's set theory, we must substitute Frege's higher-order logic.

Let us now consider Dedekind's definition of natural number. He begins (op. cit. VI (71) p. 67) by defining a *simply infinite* system. In fact N is said to be simply infinite if there exists:

> "a transformation φ of N and an element 1 which satisfy the following conditions $\alpha, \beta, \gamma, \delta$:
>
> α. $N' \ni N$
>
> β. $N = 1_0$
>
> γ. The element 1 is not contained in N'
>
> δ. The transformation φ is similar."

Dedekind continues (op. cit. VI (73) p. 68):

> "Definition. If in the consideration of a simply infinite system N set in order by a transformation φ we entirely neglect the special character of the elements; simply retaining their distinguisability and taking into account only the relation to one another in which they are placed by the ordersetting transformation φ, then are these elements called *natural numbers* or *ordinal numbers* or simply *numbers*, and the base element 1 is called the *base-number* of the *number-series* N. With reference to this freeing the elements from every other content (abstraction) we are justified in calling numbers a free creation of the human mind. The relations or laws which are derived entirely from the conditions $\alpha, \beta, \gamma, \delta$ in (71) and therefore are always the same in all ordered simply infinite systems, whatever names may happen to be given to the individual elements ... form the first object of the *science of numbers* or arithmetic."

It will be seen that Dedekind does not give a completely explicit definition of natural number; rather he gives a set theoretic structure which has the essential properties of the natural numbers, and then obtains the natural numbers by a mental abstraction. This in his view makes the natural numbers "a free creation of the human mind". Dedekind had adopted exactly the same procedure when introducing real numbers. He first defines a cut in the rational numbers, and then goes from cut to real number by mental abstraction. Russell called this procedure 'the method of "postulating" ', and ridiculed it in the following famous passage, (1919) Introduction to Mathematical Philosophy Ch. VII p. 71:

> "The method of "postulating" what we want has many advantages; they are the same as the advantages of theft over honest toil.[1] Let us leave them to others and proceed with our honest toil."

Russell's "honest toil" consists of giving explicit definitions in terms of set. Thus, for example, a real number could be defined as the set of rationals which constitute the lower half of the corresponding cut. The number 1 could be defined as the set of all sets which are similar to some particular unit set, and so on. As we have seen, Frege did give such definitions in terms of the notions "concept" and "extension of a concept. However Frege's definitions are easily reformulated in terms of sets.

Frege, as we have seen, states some principles in (1884) Foundations of Arithmetic § 78 pp. 91e-92e which are analogous to Dedekind's conditions α, β, γ, δ. For example, Frege states that 'm is a successor of n' is a 1-1 relation, which corresponds to Dedekind's condition δ, and also that every number except 0 is a successor of a number, which contains Dedekind's condition γ (with 0 replacing 1, since Frege begins the sequence of natural numbers with 0). However the logical position of these principles in Frege's account is rather different. Frege has explicitly defined natural number, successor, 0, 1; and so the principles involving these become propositions which he has to prove. If we adopted Russell's suggestion of defining numbers in terms of sets, we could then prove that the natural numbers are a simply infinite system — thereby, in a sense, integrating the approaches of Frege and Dedekind.

Once Dedekind has introduced natural numbers, the principle of mathematical induction for such numbers follows immediately from his more general theorem 59 (theorem of complete induction). Dedekind states this result as theorem 80 (op. cit. VI p. 69).

The next section (VII) of the monograph is concerned with the relations of greater and less as applied to natural numbers. Dedekind

---

[1] This incidentally was written when Russell was in prison.

has of course to define these notions in set-theoretic terms which he does as follows (op. cit. VII (89) p. 73):

> "The number m is said to be *less* than the number n and at the same time n greater than m, in symbols
>
> $m < n$, $n > m$
>
> when the condition
>
> $n_0 \ni m'_0$
>
> is fulfilled, ..."

So $m < n$, if the set of n and its successors is contained in the set of m + 1 and its successors. Dedekind goes on in the rest of the section to demonstrate the fundamental order properties of the natural numbers on the basis of this definition.

The next section (VIII) of the monograph applies the newly introduced order-concepts to a consideration of finite and infinite parts of the number-series. The main result is the following (op. cit. VIII (123) p. 83):

> "... any part T of the number-series N is finite or simply infinite, according as a greatest number exists or does not exist in T."

Dedekind's next task is to introduce addition ($m + n$), multiplication ($m.n$), and involution (a^n) of natural numbers. He does this by a method which is original to him but has since become standard; that is the method of *definition by induction*. Dedekind first gives a general theorem which justifies such definitions, and then applies it in turn to addition, multiplication, and involution. If we give the specific definitions first, however, it will make the sense of the general theorem clearer. Addition, multiplication, and involution are defined successively by induction as follows:

Addition (op. cit. XI (135) p. 97)

> "... the sum of the numbers m, n ... is completely determined by the conditions
>
> II. $m + 1 = m'$
>
> III. $m + n' = (m + n)'$ "

We are here assuming that the successor n' of any number n is given, and are defining $m + n$ in terms of this notion. The idea is to define first $m + 1$, then $m + 2$, and so on until we reach $m + n$.

Multiplication (op. cit. XII (147) p. 101)

> "... the product of the numbers m, n ... is completely determined by the conditions
>
> II. $m.1 = m$
>
> III. $m\,n' = mn + m$ "

Here we define mn using the notion of successor, and that of $m + n$ which we suppose introduced by the preceding definition.

Involution (op. cit. XIII (155) p. 104)

> "... this notion is completely determined by the conditions
>
> II. $a^1 = a$
>
> III. $a^{n'} = a.a^n = a^n a$ "

In order to justify these definitions, it is necessary to abstract their general form, and then show that such a schema defines one and only one function. This is the essential content of Dedekind's theorem (126) (op. cit. IX pp. 85-86):

> "126. Theorem of the definition by induction. If there is given an arbitrary (similar or dissimilar) transformation θ of a system Ω in itself, and besides a determinate element ω in Ω, then there exists one and only one transformation Ψ of the number-series N, which satisfies the conditions
>
> I. $\Psi(N) \ni \Omega$
>
> II. $\psi(1) = \omega$
>
> III. $\psi(n') = \theta\psi(n)$, where n represents every number."

Dedekind proves this theorem using the principle of mathematical induction which he has already established. There is however an interesting difference between his theorems regarding complete induction, and his theorem regarding definition by induction. This is discussed in a remark (op. cit. (130) pp. 88-90).

Dedekind, as we have seen, proved a generalized theorem of complete induction (his theorem) 59)) which applied to any A_0. If we specialize to the case $N = 1_0$, we get the ordinary principle of complete (or mathematical) induction (his theorem (80)). The theorem of the definition by induction (no. 126) is given only for the ordinary number-series N. Dedekind's remark is that we cannot generalize this theorem to arbitrary chains. The difficulty is that the conditions (II) and (III) of theorem (126) may, in the case of an arbitrary chain, be inconsistent. Dedekind gives the following simple example to illustrate this possibility. Let $\Omega = \{\alpha, \beta, \gamma\}$ where α, β, γ are all different. Let θ be the cyclic transformation

$$\theta(\alpha) = \beta, \theta(\beta) = \gamma, \theta(\gamma) = \alpha.$$

Now instead of N, let us take an arbitrary set $S = \{a, b\}$ where $a \neq b$. Let us define a transformation on S by

$$a' = b \qquad b' = a$$

so that $a_0 = b_0 = S$ Let us try to define a transformation ψ on S using the schema of theorem (126). We get

II. $\psi(a) = \alpha$

III. $\psi(n') = \theta\psi(n)$ where $n \, \varepsilon \, S$.

But these are inconsistent, since III gives

$$\psi(a') = \psi(b) = \theta\psi(a) = \theta(\alpha) = \beta;$$

and hence

$$\psi(b') = \theta\psi(b) = \theta(\beta) = \gamma.$$

But $\psi(b') = \psi(a)$, so that $\psi(a) = \alpha = \gamma$ — a contradiction. The difficulty here is that the general notion of chain includes cyclic chains for which the procedure of definition by induction breaks down. Dedekind does remark however that, even in the general case, if definition by induction defines a transformation at all, then that transformation is unique.

Dedekind's treatment of definition by induction is both original and of the highest importance. It can be considered as a forerunner of recursive function theory. Having defined addition, multiplication and involution by induction, Dedekind in each case proves the basic properties of these operations from his definition. Thus he proves the laws of commutativity and associativity first for addition and then for multiplication. He shows that multiplication is distributive with respect to addition, and demonstrates the laws: $a^{m+n} = a^m a^n$, $(a^m)^n = a^{mn}$, and $(ab)^n = a^n b^n$. The proofs are carried out in all cases by using the ordinary principle of mathematical (or complete) induction.

As well as his treatment of addition, multiplication and involution, Dedekind gives another application of his general theorem (126) of the definition by induction. We will conclude the present chapter with a brief mention of this matter.

In section X, Dedekind proves two theorems (op. cit. p. 92):

"132. Theorem. All simply infinite systems are similar to the number-series N and consequently ... also to one another." (op. cit. p. 93)

"133. Theorem. Every system which is similar to a simply infinite system ... is simply infinite."

From these two results, Dedekind draws the folowing conclusion (op. cit. (134) pp. 95-96):

"By the two preceding theorems (132), (133) ... every theorem in which we leave entirely out of consideration the special character of the elements n and discuss only such notions as arise from the arrangement φ, possesses perfectly general validity for every other simply infinite system Ω set in order by a transformation θ and its elements v, ... By these remarks, as I believe, the definition of the notion of number given in (73) is fully justified."

In other words, Dedekind believes that the propositions (α), (β), (γ), (δ) which define the notion of simply infinite system give an adequate characterization of the natural numbers. This optimistic conclusion has been undermined by some results of twentieth century logic.

Let us suppose that Dedekind's propositions (α), (β), (γ), (δ) are taken as axioms in some suitable formal system e.g. some formalized set theory. Call the resulting system **D**. Now one model of **D** will be the natural numbers in the ordinary intuitive sense (**N** say). Call any model isomorphic to **N** a standard model of **D**. Dedekind's theorems (132) and (133) appear to imply that **D** has only standard models. However, by a result of Skolem's, **D** (and indeed any formal system for arithmetic which has a standard model) has also non-standard models. A full discussion of this puzzling situation is to be found in Bell and Machover (1977) A Course in Mathematical Logic Ch. 7 § 2 pp. 318-324 to which the reader is referred. We will content ourselves with quoting some of Bell and Machover's concluding remarks (op. cit. p. 324):

"The informal characterization of the natural numbers works (or seems to work) only because it tacitly assumes that the notion *set of natural numbers* is interpreted correctly, as referring to *all* subsets of N. Thus it is not an *absolute* characterization but only *relative* to the notion of *power set* (set of all subsets) of a (possibly infinite) set. This latter notion cannot be characterized in a purely formal way; besides it is considerably more problematic than the notion of natural number."

The essential point seems to be this. Dedekind tries to characterize natural numbers using the notions of set theory; but these set-theoretic notions themselves cannot be unambiguously characterized in a formal way.

Chapter 10.
Peano's Axioms.
General Comparison of Frege, Dedekind, and Peano

In his interesting article already cited, Wang remarks that, (1957) The Axiomatization of Arithmetic p. 149:

> "Historically, Peano borrowed his axioms from Dedekind..."

In a way this is true, but it is unfair to Peano in as much as it suggests that Peano contributed nothing to the subject. In fact, as we shall see, Peano had a different position on the foundations of arithmetic from both Dedekind and Frege.

In his *Arithmetices principia nova methodo exposita* of 1889, Peano says in the preface (Kennedy translation p. 103):

> "In the proofs of arithmetic I used the book of H. Grassmann, *Lehrbuch der Arithmetik* (Berlin, 1861). Also quite useful to me was the recent work by R. Dedekind, *Was sind und was sollen die Zahlen* (Braunschweig, 1888), in which questions pertaining to the foundations of numbers are acutely examined."

Thus Dedekind certainly influenced Peano. How then did the two men differ? The main point is that Peano was not a logicist. Like Dedekind, but unlike Frege, Peano did admit 'class' as a logical notion (op. cit. p. 107). However he did not think that number could be defined in terms of logical notions. For Peano, arithmetic contained a number of primitive notions which could not be defined, but which could be characterized axiomatically. Thus I think it is correct to speak of Peano's axioms rather than Dedekind's axioms; for Dedekind was not trying to axiomatize arithmetic, but rather to define arithmetical notions in terms of logical ones. Another way of putting it is to say that Peano is not a logicist, but a forerunner of Hilbert's later formalism.

Peano is himself quite clear on this point and writes as follows (op. cit. p. 102):

> "Those arithmetical signs which may be expressed by using others along with signs of logic represent the ideas we can define. Thus I have defined every sign, if you except the four which are contained in the explanations of § 1. [These are number (positive integer), unity, successor, and equality (for numbers) — D.G.] If, as I believe, these cannot be reduced further, then the ideas expressed by them may not be defined by ideas already supposed to be known.

Propositions which are deduced from others by the operations of logic are *theorems*; those for which this is not true I have called *axioms*. There are nine axioms here (§ 1), and they express fundamental properties of the undefined signs."

Another difference between Dedekind and Peano is that Dedekind was a logicist, but didn't use formal logic; while Peano was not a logicist, but did use formal logic. Indeed Peano developed a system of formal logic in order to deduce theorems from axioms. His system, as a whole, is not as good as that of Frege's *Begriffsschrift*; but Peano's notation proved more acceptable than Frege's and is the ancestor of most modern systems.

Here is the passage in which Peano introduces his axioms, (1889) *Arithmetices principia*, Kennedy translation, p. 113:

"*Explanations*

The sign N means *number* (*positive integer*); 1 means *unity*; $a + 1$ means the *successor of a*, or *a plus* 1; and = means *is equal to* (this must be considered as a new sign, although it has the appearance of a sign of logic).

Axioms

1. $1 \, \varepsilon \, N$.
2. $a \, \varepsilon \, N . \supset . a = a$.
3. $a, b \, \varepsilon \, N . \supset : a = b . = . b = a$.
4. $a, b, c \, \varepsilon \, N . \supset \therefore a = b . b = c : \supset . a = c$.
5. $a = b . b \, \varepsilon \, N : \supset . a \, \varepsilon \, N$.
6. $a \, \varepsilon \, N . \supset . a + 1 \, \varepsilon \, N$,
7. $a, b \, \varepsilon \, N . \supset : a = b . = . a + 1 = b + 1$.
8. $a \, \varepsilon \, N . \supset . a + 1 -= 1$.
9. $k \, \varepsilon \, K \therefore 1 \, \varepsilon \, k . x \, \varepsilon \, k : \supset_x . x + 1 \, \varepsilon \, k :: \supset . N \supset k$.

(Here $k \, \varepsilon \, K$ means k is a class, and $N \supset k$ means N is a subset of k - D.G.)

Definitions
10. $2 = 1 + 1; 3 = 2 + 1; 4 = 3 + 1;$ etc."

Here Axioms 2, 3, 4, 5 are axioms of equality, so that Axioms 1, 6, 7, 8, 9 by themselves are often referred to as the Peano Axioms. If we write them out informally, they become:
    (P1) 1 is a number.
    (P2) The successor of any number is a number.
    (P3) Two numbers are equal if and only if their successors are equal.
    (P4) 1 is not the successor of any number.

(P5) Let k be any class. If 1 ε k, and for any number n, n ε k →
$n + 1$ ε k, then k contains the class of all numbers.
((P5) is just the principle of mathematical (or complete) induction
stated in terms of classes rather than properties).

We can see that these correspond very closely to Dedekind's conditions (α), (β), (γ), (δ) which we quoted earlier (Ch. 9 p.60 above).
Peano's (P1) is in effect stated by Dedekind in his preamble to the
conditions (α), (β), (γ), (δ). (P2), (P3), and (P4) are just different
formulations of Dedekind's (α), (δ), and (γ) respectively. Dedekind's
(β) i.e. N = 1_0, leads in the light of his Theorem 59 (Theorem of
complete induction), to the principle of complete induction for
natural numbers, given by Dedekind as theorem 80. Thus (β)
corresponds to Peano's (P5). So much then for the similarities
between Peano and Dedekind. We must next investigate more
closely the question of whether Peano's work contains a new point of
view.

I have already remarked on what seem to me two key differences
between Dedekind and Peano. First of all Dedekind seeks to define
natural number in set-theoretic terms, whereas Peano regards
natural number as an undefined notion which is characterized
axiomatically. Secondly Peano tries to develop arithmetic as a formal
system. Dedekind on the other hand has an informal approach.

It is interesting to note, however, that Peano in a paper of 1891
denies that there is any significant difference between his own
approach and Dedekind's. He writes, (1891) *Sul concetto di numero*
p. 88:

"Between what goes before, and what DEDEKIND says, there is an apparent
contradiction, which it is necessary immediately to remove. Here number is not
defined, but its fundamental properties are stated. Instead DEDEKIND defines
number, and calls number precisely that which satisfies the aforesaid conditions.
Evidently the two things coincide."

It is not however so evident to me that the two things coincide. As
we have already argued, Dedekind's approach leads naturally to
axiomatic set theory, and to an explicit definition of number and
development of number theory within set theory. Peano's approach,
on the other hand, leads naturally to the kind of formal arithmetics
considered later by Hilbert and his school. In Peano's system, we
have only to replace classes by predicates to get a typical 1st-order
formal arithmetic. There is no natural route from Peano's
Arithmetices principia to axiomatic set theory, nor, conversely, is
there any natural route from Dedekind's (1888) *Was sind und was
sollen die Zahlen?* to 1st-order formal arithmetics. If we look at the

respective positions of Peano and Dedekind in terms of their historical development, we see that there really is a difference between them.

Peano can best be considered as a forerunner of the formalist philosophy of mathematics. His *Arithmetices principia* of 1889 was just one of the first stages in a project for presenting the whole of mathematics as a vast formal system. This project was in fact carried out by Peano and his followers, and the result was the *Formulaire de Mathématiques* of which the first edition appeared in 1895 and the last in 1908. This remarkable work is written almost entirely in formulas and develops in turn: I. Mathematical Logic, II. Arithmetic, III. Algebra, IV. Geometry, V. Limits, VI. Differential Calculus, VII. Integral Calculus, and VIII. Theory of Curves. An appendix contains a few diagrams to illustrate section VIII, but this is clearly a reluctant concession since the actual treatment of section VIII involves only formulas and logical deductions.

Hilbert was undoubtedly influenced by Peano in adopting the formalist philosophy of mathematics, but, with Hilbert and his school, there is a shift of interest away from the construction of formal systems to the investigation of the metamathetical properties of such systems. In particular of course, Hilbert and his followers tried, for various formal systems, to obtain metamathematical consistency proofs which used only the methods of so-called 'finitary' arithmetic. It is interesting to note that we find the faint beginnings of this metamathematical interest in Peano.

There is no metamathematics in Peano's (1889) *Arithmetices principia,* but in his (1891) paper 'Sul concetto di numero', he gives a metamathematical investigation of the independence of the five 'Peano' axioms. This is an interesting passage, and it is worth quoting a few parts of it.

Peano begins by stating his five postulates in much the same form as we have given earlier (pp. 61-68 (P1) − (P5)). He then remarks, (1891) 'Sul concetto di numero' p. 87:

"It is easy to see that these conditions are independent."

An investigation of independence then follows on pages 87-88. Peano uses the method of models. To show that a particular postulate P say is independent of a group G say of the others, Peano devises a model M in which P is false, but the postulates of G are all true. Without giving the full investigations; we will illustrate his procedure by a couple of examples:

(i) *Proof that (P5) is independent of (P1), (P2), (P3), and (P4)* (op. cit. p. 87)

"To form a class of entities which satisfy 1, 2, 3 and 4, but not 5, it is sufficient to add to the system N another system of entities which satisfy the conditions 2, 3 and 4; thus the class formed by the positive integers N, and by the imaginary numbers of the form i + N, that is those which are obtained by adding to the imaginary unit an arbitrary positive integer, satisfy the conditions preceding 5, but not 5 itself."

(ii) *Proof that (P4) is independent of (P1), (P2), (P3), and (P5)* (op. cit. p. 87)

"To see that 4 is not even a consequence of 1, 2, 3 and 5, let us consider the roots of the equation $x^n = 1$; let us call *first root* (or 1) the imaginary root having the smallest argument $(2\pi/n)$; and let us call successor of a root a the product of a by the first root; the conditions 1, 2, 3 and 5 are verified, but not 4, the first root being also the successor of the nth. The same example can be given in popular form with the names the *hours*; one o'clock is the successor of 12 o'clock."

Such investigations of independence are very helpful because they bring to light the precise role of each of the various axioms. Although Peano investigates the independence of the axioms, he does not, so far as I have been able to discover, raise the question of consistency. It was left to Hilbert and his school to raise the question of consistency, and to investigate it in detail.

We can conveniently close this chapter by giving a table which sums up the similarities and differences between Frege, Dedekind, and Peano.

	Logicist	Psychologism in Logic	Class a Logical Notion	Formal Logic
Frege	Yes	No	No	Yes
Dedekind	Yes	Yes	Yes	No
Peano	No	—	Yes	Yes

	Philosophical Discussion	Ancestor of the following ideas
Frege	A good deal	Type Theory (Russell[1])
Dedekind	Little	Axiomatic Set Theory (Zermelo)
Peano	Little	Formal Arithmetic (Hilbert)

[1] Russell and Whitehead's *Principia Mathematica* is also influenced by Peano's notation and formalism, though its philosophical position is closer to Frege.

Chapter 11.
Frege's Begriffsschrift

Frege's *Begriffsschrift* of 1879 contains an axiomatic-dedeuctive presentation of the propositional calculus and the predicate calculus. Frege's systems are complete — though he did not prove this; and there is little in his treatment that can be faulted from a modern point of view, except for his notation which we will consider later.

Frege gives the following reason for setting up logic in an axiomatic-deductive fashion, (1879) *Begriffsschrift* §13, p. 136:

> "Because we cannot enumerate all of the boundless number of laws that can be established, we can attain completeness only by a search for those which, *potentially,* imply all the others."

This passage is interesting because Frege implies that his system is indeed complete, though he does not attempt a precise definition and proof of this. Frege calls the set of those laws which potentially imply all the others the *kernel* of his system, and he goes on to describe it as follows (op. cit. § 13 p. 136):

> "Nine propositions form the kernel of the following presentation. Three of these — formulas 1, 2, and 8 — require for their expression (except for the letters), only the symbol of conditionality. Three — formulas 28, 31, and 41 — contain in addition the symbol for negation. Two — formulas 52 and 54 — contain the symbol for identity of content; and in one — formula 58 — the concavity in the content stroke is used."

Frege's "concavity in the content stroke" is the universal quantifier.

Frege's "kernel" consists of the axioms of his logic. We will now write them out, changing Frege's symbols for connectives, quantifiers, and identity into the ones used in this book (see Appendix I On Notation), but retaining the letters used by Frege.

Axioms containing only →

(1) $a \rightarrow (b \rightarrow a)$
(2) $(c \rightarrow (b \rightarrow a)) \rightarrow ((c \rightarrow b) \rightarrow (c \rightarrow a))$
(8) $(d \rightarrow (b \rightarrow a)) \rightarrow (b \rightarrow (d \rightarrow a))$

Axioms containing both → *and* ¬
(28) $(b \rightarrow a) \rightarrow (\neg a \rightarrow \neg b)$
(31) $\neg\neg a \rightarrow a$
(41) $a \rightarrow \neg\neg a$

Axioms of identity
(42) $(c = d) \rightarrow (f(c) \rightarrow f(d))$
(54) $c = c$

Axiom of the Universal Quantifier
(58) $(\forall \mathbf{a})f(\mathbf{a}) \rightarrow f(c)$

Frege states (op. cit. § 1 p. 111) that the letters are to be considered as variables. Now in many modern presentations, the Axiom of the Universal Quantifier would be stated as

$(\forall x)f(x) \rightarrow f(y)$

where it has to be further specified that no free occurrences of x in $f(x)$ lie within the scope of a quantifier $(\forall y)$ or $(\exists y)$. This qualification is added to avoid difficulties such as the following. Let $f(x)$ be $(\exists y)$ $(y \neq x)$. Then $(\forall x)f(x)$ becomes $(\forall x)$ $(\exists y)$ $(y \neq x)$ and is true in any domain having two or more members, whereas $f(y)$ becomes $(\exists y)$ $(y \neq y)$ which is always false. Frege avoids the need for such a qualification by introducing a new type of variable (German[1] as opposed to italic letters) for quantifiers.

Frege sometimes claims to use only one rule of inference viz. *modus ponens*: from B and B → A, A follows (c.f. Frege, (1879) *Begriffsschrift* § 6 p. 117). In fact, however, he uses three others viz. substitution, generalization, and confinement. Frege constantly makes substitutions in his proofs, but he never formulates precise rules governing substitution. The rule of generalization,[2] he states as follows (op. cit. § 11 p. 132):

"... instead of X (*a*) we may put $(\exists \mathbf{a})X(\mathbf{a})$ if *a* occurs only in the argument places of X(*a*)"

The rule of confinement[2] is given as follows (op. cit. § 11 p. 132):
"*It is also obvious that from*
$A \rightarrow \Phi(a)$
we can derive
$A \rightarrow (\forall \mathbf{a})\Phi(\mathbf{a})$

---

[1] We have used bold face instead of Frege's German letters.
[2] I have changed Frege's notation as in the axioms.

if A *is an expression in which a does not occur and a stands only in the argument places of* Φ (*a*)." (Frege's italics)

While Frege in places seems to ignore the rules of inference other than *modus ponens,* elsewhere he is more careful. Thus he writes, (1879) *Begriffsschrift* § 6 p. 119:

> "In logic people enumerate, following Aristotle, a whole series of modes of inference. I use just this one (i.e. *modus ponens* – D.G.) – *at least in all cases where a new judgement is derived from more than one judgement."* (my italics)

The qualification in italics makes what Frege says here correct – though he does not fully clarify the matter.

Frege's first six axioms, together with the rules of *modus ponens* and substitution give a complete system for the propositional calculus. However the axioms are not independent. Łukasiewicz showed that the third axiom can be deduced from the first two (c.f. Łukasiewicz (1934) On the History of the Logic of Propositions pp. 86-7, where the formal derivation is given). A simple and attractive axiom scheme for the propositional calculus can be obtained by retaining Frege's first two axioms (Formulas (1) and (2)), and replacing his next four axioms (Formulas (8), (28), (31), and (41)) by a single axiom viz.

$$(\neg a \rightarrow b) \rightarrow ((\neg a \rightarrow \neg b) \rightarrow a)$$

which can be thought of as expressing a form of *reductio ad absurdum.* If we take *modus ponens* as a rule of inference, and regard the axioms as axiom-schemas (or alternatively add a rule of substitution), the resulting system is consistent and complete. (For details see Bell and Machover (1977) Ch. 1 § 10 f.)

Frege intended his system as a higher-order logic, that is, he allowed quantification over predicates, and in fact does quantify over predicates in several formulas of the *Begriffsschrift* (e.g. Formula (76)). However an appropriate fragment of his system can be interpreted as a system of 1st-order predicate calculus with identity, and, if so interpreted, is complete. Thus, if we put aside the question of notation, it must be said that modern syntactic presentations of the propositional and predicate calculus are very little superior to Frege's *Begriffsschrift* treatment.

The most noted 19th century work on formal logic before Frege was of course Boole's (1847) The Mathematical Analysis of Logic. Between 1847 and 1879 most researchers interested in formal logic worked on extending and improving Boole's system. Frege's work, however, contained many remarkable innovations and improvements vis-à-vis this Boolean tradition. His axiomatic-deductive presentation of the propositional calculus (with axioms and rules of inference clearly distinguished), and his complete development of predicate calculus and quantification theory are all novel. The question naturally arises: what enabled Frege to make such striking advances?[1]

The key difference between the systems of Boole and Frege can best be understood if we take account of their different approachs to logic and different motivations for studying the subject. Let us start with Boole. He belonged to the British school of algebra which flourished in the 19th century. The members of this school developed new algebraic techniques and systems, and applied these to a variety of problems in mathematics and physics. Boole's first piece of mathematical research his 1844 'A General Method in Analysis' applies this kind of approach to analysis by developing a calculus of operators. His next idea was to deal with logic in the same way — that is to say to reduce the methods of traditional logic to an algebraic calculus. The title of Boole's 1847 work: 'The Mathematical Analysis of Logic, being an essay towards a calculus of deductive reasoning', clearly indicates Boole's programme. In the first chapter of this work 'First Principles', Boole sets out the basic operations of his algebraic calculus. He begins each of the next four chapters, which constitute the bulk of the monograph, by summarizing some of the basic doctrines of traditional logic. He then goes on to show how these doctrines can be expressed using his algebraic calculus. Thus Boole's programme is really to reduce traditional logic to algebraic formulas and manipulations. Of course his new notation and approach do suggest extensions of traditional logic at various points, but there is nothing in the programme likely to bring about a dramatic alteration in the content of traditional logic.

Let us next contrast Boole's programme with Frege's. Frege's principal aim was to establish his logicist view that arithmetic could be reduced to logic. To this end he had to develop arithmetic as an axiomatic-deductive system, and to show that all the axioms needed were truths of logic. He had moreover to make all the rules of inference used in the proofs fully explicit in order to make sure that no inferences were used which depended on some kind of intuition rather than on pure logic. These aims Frege states very clearly in the Preface to the *Begriffsschrift* (p. 104):

[1] On this question of the relationship between Boole and Frege, I have greatly benefited from long conversations with Dr. M. L. G. Redhead. Indeed the views to be expressed on this matter are largely due to him.

"Now, while considering the question to which of these two kinds [of truths] do judgements of arithmetic belong, I had first to test how far one could get in arithmetic by means of logical deductions alone, supported only by the laws of thought, which transcend all particulars. The procedure in this effort was this: I sought first to reduce the concept of ordering-in-a-sequence to the notion of *logical* ordering, in order to advance from here to the concept of number. So that something intuitive could not squeeze in unnoticed here, it was most important to keep the chain of reasoning free of gaps. As I endeavoured to fulfil this requirement most rigorously, I found an obstacle in the inadequacy of the language; despite all the unwieldiness of the expressions, the more complex the relations became, the less precision — which my purpose required — could be obtained. From this deficiency arose the idea of the "conceptual notation" presented here. Thus, its chief purpose should be to test in the most reliable manner the validity of a chain of reasoning and expose each presupposition which tends to creep in unnoticed, so that its source can be investigated."

The difference between Frege's programme and Boole's also shows up in the differences between their respective monographs. Frege, unlike Boole, hardly considers traditional logic (except in passing). Having set up the propositional and predicate calculi in Chs. I and II, Frege proceeds in Ch. III to 'Some topics from a general theory of sequences'. This investigation, as we have seen, forms part of his attempt to reduce mathematical induction (or the inference from n to $n + 1$) to purely logical inference.

Thus whereas Boole wanted to express traditional logic more perspicuously using the techniques of algebra, Frege wanted to distill out the logic needed to develop arithmetic deductively. Frege states that in his system: "calculation becomes deduction" ((1884) Foundations of Arithmetic § 87, p. 99ᵉ), while Boole aimed to produce: "a calculus of deductive reasoning" (part of the subtitle of his (1847) The Mathematical Analysis of Logic) i.e. to reduce deduction to calculation. We could put the difference aphoristically as follows.[1] Boole tried to reduce logic to arithmetic, and Frege to reduce arithmetic to logic. The meaning here is that Boole wanted to reduce deductive logic to an algebraic calculus similar to the algebraic calculus abstracted from the usual arithmetical operations. In this sense he aimed to reduce logic to arithmetic.

We have next to show why Frege's programme produced greater innovations in Logic than Boole's. The point is really quite a simple one. Frege had to make fully explicit all the logical principles needed in a deductive development of arithmetic, and in fact many of these go beyond anything which had been recognized in traditional logic. On the other hand, there was, as we have already argued, little reason

[1] This formulation is due to Dr. M. L. G. Redhead.

why Boole's programme should lead to striking changes in the content of traditional logic.

We can amplify this by looking at the innovations of Frege's mentioned earlier. First of all an axiomatic-deductive presentation of logic would clearly be needed for Frege's programme — though not for Boole's. Secondly an adequate treatment of quantification theory is necessary for formalizing arithmetic. Consider the simple statement, mentioned earlier, that there is a prime number greater than any given number. If we write: 'm is a prime number' as Pr (m), this becomes

$$(\forall n)(\exists m)(\Pr(m) \wedge (m > n)).$$

However here we have the nested quantifiers $(\forall n)(\exists m)$, and this goes beyond anything to be found in traditional logic. Then again to express the principle of mathematical induction in its first-order form:

$$(P(0) \wedge (\forall n)(P(n) \to P(n+1))) \to (\forall n)P(n)$$

we have to have the notion of the scope of a quantifier. This notion is in fact introduced by Frege in (1879) *Begriffsschrift* § 11 p. 131, and is used in his definition of 'the property F is hereditary in the φ-sequence'. If we write 'y is a successor of x in the φ-sequence' as '$\varphi(x,y)$', then Frege's definition in modern notation becomes:

$$(\forall x)(F(x) \to (\forall y)(\varphi(x,y) \to F(y))$$

(c.f. (1879) *Begriffsschrift* § 24 pp. 167-170). This definition plays an essential part in Frege's attempt to reduce mathematical induction to logical inference, as we remarked earlier.

So Frege's logicist programme provided the stimulus for his advances in formal logic, but, if the matter is considered carefully, it will I think be seen that only a part of this programme would have been sufficient by itself to provide the necessary stimulus. The crucial thing is the plan to develop arithmetic as a formal axiomatic-deductive system, i.e. as an axiomatic-deductive system in which the underlying logic is made fully explicit. This view will, I believe, be supported by a consideration of Peano's work on logic and the foundations of arithmetic.

As we have seen, in his *Arithmetices principia nova methodo exposita* of 1889, Peano tries to give a formal axiomatic-deductive development of arithmetic. However Peano, unlike Frege, is not a logicist. He believes that arithmetic contains certain primitive notions which cannot be reduced to logical notions; and also depends on certain arithmetical axioms which cannot be derived from logical axioms.

It is indicative of Frege's failure to gain recognition that Peano writing 10 years after the *Begriffsschrift* does not refer to Frege at all, and almost certainly had not heard of him. If our general thesis is correct, Peano's project for the foundations of arithmetic should have stimulated him, like Frege before him, to make advances in logic. This turns out to be the case, except that, whereas Frege developed quantification theory in a complete form which has hardly been improved on since, Peano's work was much more scrappy and stood in need of a great deal of development and improvement. Still Peano was forced by the requirements of his programme for arithmetic to find some means of expressing what is now expressed using the universal and existential quantifiers. Let us next examine what he did.

The first relevant passage is the following, (1889) *Arithmetices principia nova methodo exposita* p. 105:

> "If the propositions *a, b* contain the indeterminate quantities *x, y, ...*, that is, express conditions on these objects, then $a \supset_{x, y...} b$ means: whatever the *x, y, ...*, from proposition *a* one deduces *b*. If indeed there is no danger of ambiguity, instead of $\supset_{x, y...}$ we write only $\supset$."

This device enables Peano to express propositions like $(\forall x)(A(x) \rightarrow B(x))$, $(\forall x)(\forall y)(A(x, y) \rightarrow B(x, y))$, ... However he cannot use it to express propositions like $(\forall x)A(x)$, $(\exists x)(A(x) \rightarrow B(x))$, or $(\forall x)(\exists y)A(x, y)$. To increase the expressive power of his symbolism, Peano has to make use of his notation for classes. Peano regarded 'class' as a logical notion, and developed a calculus of classes as part of his logic. In particular he writes (op. cit. p. 108):

> "Let *a* be a proposition containing the indeterminate *x*; then the expression [*xε*] *a*, which is read *those x such that a*, or *solutions*, or *roots* of the condition *a*, indicates the class consisting of individuals which satisfy the condition *a*."

Note that here Peano implicitly assumes the so-called axiom of comprehension i.e. $(\exists y)(\forall x)(x \; \varepsilon \; y \leftrightarrow P(x))$ which leads to Russell's paradox (see Ch. 12) below .

Using this device of class abstraction, Peano can express existential quantification. For example $(\exists x)a(x)$, he could write as $[x\varepsilon].a: -= \Lambda$, where Λ is the null class, and '$-=$' means 'is not equal to'. Indeed Peano has to make use of this device in order to express some of the theorems in his subsequent development of arithmetic. For example, his § 8 Theorem 12 (op. cit. p. 126) would be written using the standard quantifiers as:

$$(\forall p, q)(p, q \; \varepsilon \; N. \supset (\exists m)(m^p/_q \; \varepsilon \; N).$$

Peano writes it as:

$$p, q \; \varepsilon \; N. \supset :: [m \; \varepsilon] : m \; \varepsilon \; N. m^p/_q \; \varepsilon \; N \therefore -= \Lambda.$$

We cannot leave the subject of Frege's innovations in formal logic without briefly inquiring why they failed to gain recognition from Frege's contemporaries. It might be thought that Frege's work was too formal and technical in character, and was therefore left unread. This, however, cannot be the whole answer, for Peano's work was equally formal and technical, and yet gained instant recognition (and even popularity). Moreover Bynum in his recent (1972) translation of the *Begriffsschrift* has conveniently collected the six contemporary reviews of the *Begriffsschrift*, translated into English where necessary, into Appendix I. These reviews show that Frege's work was read by some of the leading logicians of the time. The trouble was that they failed to appreciate Frege's innovations. Thus Venn writes in his review in Mind (1880) (p. 234 of Bynum's (1972) edition of the *Begriffsschrift*):

"... it does not seem to me that Dr. Frege's scheme can for a moment compare with that of Boole. I should suppose, from his making no reference whatever to the latter, that he has not seen it, nor any of the modifications of it with which we are familiar here. Certainly the merits which he claims as novel for his own method are common to every symbolic method."

In effect Venn has completely failed to see Frege's advance over Boole, though he may be right about one thing, namely that Frege had not read Boole before composing the *Begriffsschrift*.

Of course Frege is not the first thinker (and without doubt will not be the last) whose innovations are not understood by contemporaries familiar with older ways of thought. However, there is one particular factor which may have rendered his work difficult for his contemporaries, and that is his peculiar two-dimensional notation. Certainly the most detailed review of Frege (that by Schröder) singles this feature out for (often quite justified) criticism, and Frege's two-dimensional script is the one part of his logic which has never been accepted. We will now briefly explain Frege's notation, and the objections which can be raised to it.

Frege writes the content of a proposition A as

$$\text{————}A$$

If the proposition is asserted, he writes

$$\vdash\text{————}A$$

where the small vertical line is his assertion sign. He bases his treatment of the propositional calculus on two connectives, material

implication and negation (in our notation → ¬). ¬ A he writes
——┬——A, which is quite unobjectionable. However A → B, he
writes

This procedure gives Frege's *Begriffsschrift* its peculiar two dimensional character. The notation does allow us to dispense with brackets. Thus A → (B → C) is written

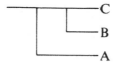

while (A → B) → C is written

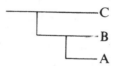

Frege's second axiom for the propositional calculus, which, in our notation, is

$$(c → (b → a)) → ((c → b) → (c → a))$$

is written by him, (1879) Begriffsschrift § 15 p. 140:

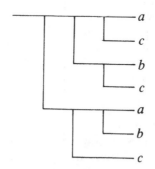

This gives a vivid illustration of how Frege's notation converts a horizontal row into a vertical column

Schröder comments as follows on this notation of Frege's, (1880) Review of *Begriffsschrift* p. 229:

"In fact, the author's formula language not only indulges in the Japanese practice of writing vertically, but also restricts him to only *one* row per page, or at most, if we count the column added as explanation, two rows! This monstrous waste of space which, from a typographical point of view (as is evident here), is inherent in the Fregean "conceptual notation", should definitely decide the issue in favour of the Boolean school — if, indeed, there is still a question of choice."

Those accustomed to reading European languages do indeed find it easier to follow a script written in rows from left to right. In English, for example, the conditional is written:

If A, then B

so that the symbolic

A → B

which bears an obvious analogy is easy to understand. Frege's

is correspondingly difficult to grasp. Of course it may be that Frege's notation is easier for those accustomed to read Chinese or Japanese — but this is something on which I cannot comment.

Schröder is also right when he says that Frege's notation is a "waste of space", as the reader may easily see by comparing Frege's second axiom written in his own notation, with the same axiom written in the more usual notation (see above p. 79). However Schröder is guilty of a *non-sequitur* when he says that this waste of space (op. cit. p. 229):

"... should definitely decide the issue in favour of the Boolean school..."

At most the waste of space shows that Frege has a bad notation for the material conditional — not that his system as a whole is inferior to Boole's. However this *non-sequitur* may have been at least partly responsible for the poor reception of the *Begriffsschrift*.

Another disadvantage of Frege's notation is that it does not allow us to introduce abbreviations for the other connectives. Suppose, for example, we give an axiomatic-deductive development of the propositional calculus, introducing →, ¬ as the primitive connectives. We do not, of course, have to introduce any further connectives, since →, ¬ by themselves suffice to express any compound proposition of the calculus. However it is nonetheless very convenient for clarity and conciseness to introduce the other connectives as abbreviations e.g.

A v B = _{def}¬A → B
A ∧ B = _{def}¬ (A →¬B) etc.

But Frege's notation does not allow us to introduce such abbreviations in any convenient way. Thus all compound propositions must be written out in the primitive notation. Schröder points out one particularly striking instance of this (op. cit. p. 227):

"Now, in order to represent for example the disjunctive "or" – namely, to state that *a holds or b holds, but not both* – the author has to use the schema

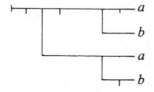

Admittedly the disjunctive "or" is not usually employed, but, should it be needed, we can in the standard treatment easily introduce an abbreviation (A⊽ B is sometimes used for this). However, in Frege's notation, the complicated expression just given has to be written out each time.

It should be observed, incidentally, that Schröder had obviously read the *Begriffsschrift* carefully. He points out a mistake which Frege made when he wrote, (1879) *Begriffsschrift* § 5, p. 117:
"We can see just as easily that

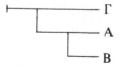

denies the case in which B is affirmed, but A and Γ are denied." In fact Frege's verbal definition corresponds to the formula B → (A∨Γ) rather than to the one he gives i.e. (in our notation) (B → A) → Γ. Schröder remarks in this context (op. cit. p. 225):
"... the author infortunately makes a mistake (p. 7 – however, it is the only one which I noticed in the whole book)..."
Schröder does also admit (op. cit. p. 229) that Frege can express generality better than Boole, but he adds (op. cit. pp. 229-30):
"... one may not perchance find a justification here for his other deviations from Boole's notation, and the analogous modification or extension can easily be achieved in Boolean notation as well."
It is possible then that if Frege had replaced his two dimensional notation for material implication by a linear one, his work might have been more favourably received. However, Frege stuck to his guns and rejected the views of his critics on this point. In a reply to

Schröder's review he wrote, (1882) On the Aim of the *Begriffsschrift* p. 97:

"The disadvantage of the waste of space of the *Begriffsschrift* is converted into the advantage of perspicuity; the advantage of terseness for Boole is transformed into the disadvantage of unintelligibility. The *Begriffsschrift* makes the most of the two-dimensionality of the writing surface by allowing the assertible contents to follow one below the other while each of these extends [separately] from left to right. Thus, the separate contents are clearly separated from each other, and yet their logical relations are easily visible at a glance. For Boole, a single line, often excessively long, would result."

Frege's reply to the Boolean Schröder is interesting because it betrays at one point a certain lack of confidence. It is true that Frege definitely claims his treatment of quantification to be an advance on Boole. Speaking of his notation for the universal quantifier, he says, (1882) On the Aim of the *Begriffsschrift* p. 99:

"I consider this mode of notation one of the most important components of my *Begriffsschrift*, through which it also has, as a mere presentation of logical forms, a considerable advantage over Boole's mode of notation."

On the other hand, when comparing his *Begriffsschrift* with the Leibnizian-Boolean formula language, he writes (op. cit. p. 98):

"We can ask ... whether perhaps my formal language governs a smaller region."

Was Frege himself at least partially unaware of the superiority of his logic to that of the Booleans?

Frege did not change his mind about his two-dimensional notation. In an article of 1896 comparing his system of formal logic with Peano's he writes, (1896) *Uber die Begriffsschrift des Heern Peano und meine eigene.* p. 222:

'In Peano's formal logic the writing of formulas on a single line is, so it seems, carried through as a fundamental principle which appears to me as a wanton renunciation of a major advantage of writing over speech. The convenience of the typesetter is not however the highest Good. For physiological reasons, a long line is harder to survey and its divisions are harder to grasp than shorter lines lying underneath each other, and created from the breaking up of the original line, provided that this partition corresponds to the division of the sense."

Chapter 12.
Frege's Grundgesetze, and Russell's paradox.

Between the *Begriffsschrift* of 1879 and the *Grundgesetze der Arithmetik* of 1893, Frege made many alterations in, and developments of, his logical system. One of the most important innovations was the introduction and elaboration of his (now famous) distinction between *sense* (*Sinn*) and *reference* or *denotation* (*Bedeutung*). Frege devoted an article, his (1892) "On Sense and Reference", to discussing this distinction, and incorporated the results in the logic of the *Grundgesetze*.

The problem is concerned with equality statements of the form $a = b$. If we take equality as a relation between the objects denoted by a and b, then, if $a = b$ is true, it would seem to express the same relation as $a = a$. Yet, Frege objects, $a = a$ is always trivial, whereas $a = b$ can sometimes contain a valuable extension of our knowledge. The next suggestion considered by Frege is that $a = b$ should be taken to mean: the name 'a' and the name 'b' stand for the same object. But now equality statements would only express linguistic knowledge about what names we have given to certain objects. However equality statements can sometimes express knowledge of other kinds. For example (Frege's own example), 'The morning star = the evening star' expresses astronomical rather than linguistic knowledge.

We are thus led to Frege's final theory. The reference or denotation of 'the morning star' and of 'the evening star' is the actual planet Venus. On the other hand the sense of the expression 'the morning star' differs from the sense of the expression 'the evening star', and this is why the proposition 'the evening star = the morning star' can express significant knowledge. Frege sometimes speaks of the sense of a referring expression as the mode of presentation of the object. The idea here is presumably that the planet Venus can be presented either as the star which shines brightly in the early morning, or as the star which shines brightly in the early evening. In general, Frege analyses "$a = b$" as "the sense of 'a' and the sense of 'b' have the same reference".

It is worth noting the connection between Frege's logical distinction between sense and reference, and his metaphysics. Frege always accepted the existence of a real, external, material world, and he thought of certain linguistic expressions as being used to refer to objects in this world. Russell, on the other hand, wanted to reduce material objects to collections of sense-data, and this was part of the reason for his rejecting the sense/reference distinction. (c.f. Russell (1905) On Denoting).

Another link with Frege's metaphysics is formed when Frege extends the sense/reference distinction to mathematics. He analyses the equality '$2 + 2 = 3 + 1$' along the same lines as the equality 'the morning star = the evening star'. The expressions '$2 + 2$' and '$3 + 1$' have different senses, but the same reference viz. the number 4. This analysis presupposes Frege's Platonic view that there is a 'third world' of objectively existing abstract entities (e.g. numbers, propositions, etc.) analogous to the 'first world' of material objects. Incidentally Frege's analysis is still acceptable if we replace his 'traditional Platonism' by 'constructive Platonism,' as suggested in Ch. 6.

In the mathematical case, Frege gives an interesting example to illustrate his view of sense as 'mode of presentation', (1892) On Sense and Reference, p. 57:

> "Let a, b, c, be the lines connecting the vertices of a triangle with the midpoints of the opposite sides. The point of intersection of a and b is then the same as the point of intersection of b and c. So we have different designations for the same point, and these names ('point of intersection of a and b', 'point of intersection of b and c') likewise indicate the mode of presentation; and hence the statement contains actual knowledge."

As far as denoting expressions of the form 'the morning star', 'the even prime' etc. are concerned, Frege's distinction between sense and reference seems to me most valuable, and it does help to clarify the meaning of equality statements of the form '$a = b$'. However Frege proceeds to extend the distinction to whole sentences. This extension appears to me to be of much more questionable value.

Let us take as an example the sentence '$2 + 2 = 4$.' The sense of this sentence is, for Frege, the proposition (or thought in the objective sense) expressed. Such propositions or thoughts are inhabitants of his third world. So far, the extension seems reasonable, but Frege next argues that we should consider the truth-value of the proposition expressed as the reference of the sentence. Thus the reference of '2 + 2 = 4' would be the truth-value: the True. But this seems to me metaphorical and strained. Is it really reasonable to regard '2 + 2 = 4' as referring to the True, in the same way as 'the morning star' refers to the planet Venus? I would say: no. This view of truth-values, however, is embodied in the logic of the *Grundgesetze* to which we now turn.

In the *Begriffsschrift*, as we have seen, Frege wrote the content of a proposition A as

$$\underline{\hspace{3cm}} A$$

and referred to —— as the *content-stroke*. Having separated the content into thought and truth-value, he now uses

$$\underline{\hspace{3cm}} A$$

to denote the truth-value of A, and refers to —— simply as the *horizontal*. Accordingly we have the following explanation, (1893) *Grundgesetze* Vol I § 5 Furth Translation p. 38:

"
$$\underline{\hspace{3cm}} \Delta$$

is the True if Δ is the True; on the other hand it is the False if Δ is not the True."

One point to note about this exaplanation is that we do not have to restrict ——Δ to cases where Δ expresses a proposition. We can substitute for Δ any referring expression, and write —— the Queen of England in 1879, or —— 2. Since neither the Queen of England in 1879 nor 2 is the True, both these expressions stand for the False.

Correspondingly, In Frege's sign for the conditional,

$$\begin{array}{l} \underline{\hspace{2cm}} \Gamma \\ \underline{} \Delta \end{array}$$

we do not have to restrict Γ, Δ to signs expressing propositions, but can write e.g.

$$\begin{array}{l} \underline{\hspace{2cm}} \text{the Sun} \\ \underline{} 3 \end{array}$$

This actually turns out to be true, since —— the Sun and —— 3 both stand for the False, and it is true that the False materially implies the False.

This curious approach of Frege's makes it difficult to translate the notation of the *Grundgesetze* accurately into a more modern notation. If, as in the case of the *Begriffsschrift* we translate

85

$$\begin{array}{r}\overline{\hspace{2cm}}\raisebox{0.5em}{\rule[0.5em]{0.01em}{1.2em}}\Gamma\\ \hspace{0.6cm}\Delta\end{array}$$

as

$$\Delta \rightarrow \Gamma \qquad ,$$

we are not doing justice to Frege, since e.g. '2 → the Sun' would normally be regarded as nonsense. Nonetheless it still seems to me worthwhile, in the interests of claity, to translate Frege's formulas into a modern notation. We are thereby restricting the meaning of Frege's logic, but, if this is clearly understood, it should cause no harm, since the reader can always mentally supply the extension should he wish to do so.

Our procedure then will be to translate Frege's notation for connectives, quantifiers, extensions of concepts, membership etc. into the modern notation given in our appendix. As in the previous chapter, we will retain Frege's letters for variables, functions etc. except that we will replace the small Greek 'ε' wherever it occurs by 'θ'. This is to avoid confusion with the membership symbol. In several further cases (e.g. the notion of extension of a concept), this translation will involve a restriction of Frege's original meaning, and we will draw attention to some such restrictions as we go along.

We have objected to the view that a proposition refers to its truth-value. Still the notion of truth-value (which Frege was the first to introduce) is a most valuable one. In his (1879) *Begriffsschrift*, § 5, Bynum translation pp. 114-5, Frege explained the meaning of his conditional stroke as follows:

"If A and B stand for assertible contents (§ 2), there are the following four possibilities:
    (1) A is affirmed and B is affirmed
    (2) A is affirmed and B is denied
    (3) A is denied and B is affirmed
    (4) A is denied and B is denied.
Now,

stands for the judgement that *the third of these possibilities does not occur, but one of the other three does.*"

This explanation in terms of affirming and denying contains a whiff of psychologism. It is therefore a considerable improvement from Frege's point-of-view to replace it by an explanation in terms of truth-values. This Frege does in his (1893) *Grundgesetze,* § 12, Furth translation p. 51, as follows:

"... I introduce the function of two arguments

by stipulating that its value shall be the False if the True be taken as ζ-argument and any object other than the True be taken as ξ-argument, and that in all other cases the value of the function shall be the True."

Frege formulates the propositional and predicate calculi in a different way in the *Grundgesetze* from the one he gives in the *Begriffsschrift*. The essential difference is that in the *Grundgesetze* he has fewer axioms and more rules of inference. The reson for this change is that it enables Frege to shorten some of the formal proofs. As Frege himself says, (1893) *Grundgesetze,* Vol. I, § 14, Furth translation p. 57:

"This (i.e. Modus ponens - D.G.)is the sole method of inference used in my book *Begriffsschrift,* and one can actually manage with it alone. The dictates of scientific economy would properly require that we do so; yet in this book, where I wish to set up lengthy chains of inference, I must make some concessions to practical considerations. In fact, if I were not willing to admit some additional methods of inference the result would be exorbitant lengthiness — a point already anticipated in the Foreword to *Begriffsschrift.*"

We will not expound Frege's new version of the propositional and predicate calculi, since the changes from the *Begriffsschrift* are motivated by practical considerations concerned with the ease of technical manipulation. We will concentrate rather on those parts of the logic of the *Grundgesetze* where Frege introduces new principles having some theoretical importance. One such theoretical change is that the logic of the *Grundgesetze* is more explicitly a higher-order logic. This is not something wholly new, since the *Begriffsschrift* was always intended to be a higher-order logic, and indeed, as we have seen, Frege does quantify over predicates in the *Begriffsschrift*. Still, the whole matter is made more precise and explicit in the *Grundgesetze*.

The Basic Law (Axiom) IIa of the *Grundgesetze* is

$(\forall a)f(a) \rightarrow f(a)$

which is the same as the *Begriffsschrift's* Axiom of the Universal Quantifier. However, in the *Grundgesetze*, Frege explicitly extends this to deal with the case of quantifying over functions. To do so, he has to clarify somewhat the notion of function.

Frege assumes a domain of objects which includes both material objects, and abstract objects. He gives the following explanation, (1893) *Grundgesetze,* § 21, Furth translation, p. 74:

"We now call those functions whose arguments are objects *first-level functions*; on the other hand, those functions whose arguments are first-level functions may be called *second-level functions.*"

A first-level function of one argument is called an argument of type 2.

Frege's next step is to introduce a notation for second-level functions, (1893) *Grundgesetze,* § 25, Furth translation, pp. 79-80:

"We indicate a second-level function of one argument of type 2 in this way:
"$M_\beta(\varphi(\beta))$"
by using the *Roman function-letter* "*M*", as we indicate a first-level function of one argument by "$f(\xi)$". "$\varphi(\quad)$" here renders recognizable the argument-place, just as "ξ" does in "$f(\xi)$". The letter "β" here in the brackets fills up the place of the argument of the function occurring as argument [of the whole]."

Using this notation, Frege states his basic law IIb as follows:

$(\forall f)M_\beta(\mathbf{f}(\beta)) \to M_\beta(f(\beta))$

This, of course, is precisely analogous to IIa, except that the quantification is over first-level functions of one argument rather than over objects.

Frege also gives a higher-order treatment of equality (or identity). The axioms of identity given in the *Begriffsschrift* are replaced by the *Grundgesetze's* basic law III:

$g(a = b) \to g(\,(\forall f)(\mathbf{f}(b) \to \mathbf{f}(a)))$

If '$g(...)$' is replaced successively by '... is true' and '... is false', the two *Begriffsschrift* axioms of identity viz.

$(a = b) \to (f(a) \to f(b))$
$a = a$

are easily derived. Frege carries out the derivations in (1893) *Grundgesetze,* Vol. I, § 50, Furth translation. pp. 111-113.

So far then Frege has either reformulated parts of the *Begriffsschrift* (in his treatment of the propositional and predicate calculi), or made more explicit what was implicit (in his treatment of higher-order quantification). Next, however, he introduces something which is quite novel *vis-à-vis* the *Begriffsschrift* and which was to lead to contradiction. This is his formal treatment of the extensions of concepts. In his (1884) Foundations of Arithmetic, Frege had defined numbers as extensions of concepts. His logicist programme therefore required him to add to his logic a section dealing with extensions of concepts.

Consider a concept Φ (...), Frege writes the extension of the concept, Φ (...), or the set of all θ which satisfy Φ (...), as $\theta\Phi(\theta)$. This is quite similar to the now more usual $\theta\Phi(\theta)$ which we shall use to translate his notation. However, as in the case of the conditional, Frege uses $\theta\Phi(\theta)$ in an extended sense. He allows one to substitute for

Φ (...) not just a concept, but any function of one argument, so th.
one can speak e.g. of $\theta(\theta^2-\theta)$. Indeed he speaks of $\grave{\theta}\Phi(\theta)$ as tl
course-of-values of the function Φ (...). Frege in the *Grundgeset*
regards a concept as a function whose possible values are the Tr
and the False. Thus the extension of a concept is a special case of the
course-of-values of a function. When we translate Frege's $\grave{\theta}\Phi(\theta)$ as
$\theta\Phi(\theta)$, it should be remembered that we are once again somewhat
restricting Frege's meaning.

Frege introduces his basic law V to deal with the extensions of
concepts. It is:

$$(\grave{\theta}f(\theta) = \hat{\alpha}g(\alpha)) \leftrightarrow (\forall a)(f(a) \leftrightarrow g(a))$$

This was the law which led to contradiction, and it is interesting to
note that even before the discovery of the contradiction, Frege expressed some doubts about his basic law V. In the introduction to the
(1893) *Grundgesetze*, he writes (Furth Translation pp. 3-4):

> "If anyone should find anything defective, he must be able to state precisely where,
> according to him, the error lies: in the Basic Laws, in the Definitions, in the Rules,
> or in the application of the Rules at a definite point. If we find everything in order,
> then we have accurate knowledge of the grounds upon which each individual
> theorem is based. A dispute can arise, so far as I can see, only with regard to my
> Basic Law concerning courses-of-values (V), which logicians perhaps have not yet
> expressly enunciated, and yet is what people have in mind, for example, where
> they speak of the extensions of concepts. I hold that it is a law of pure logic. In any
> event the place is pointed out where the decision must be made."

Frege also introduces a sign which serves as a substitute for the
definite article of everyday language. Using the more modern
symbol $\iota\theta\Phi(\theta)$ (i.e. the θ such that $\Phi(\theta)$), his explanation is as follows.
If Φ (...) is a concept under which one and only one object Δ falls,
then $\theta\Phi(\theta) = \Delta$. Otherwise $\iota\theta\Phi(\theta) = \grave{\theta}\Phi(\theta)$. Actually Frege's own
explanation is again more general, though we will not go into details.

In his (1905) article "On Denoting", Russell points out a number
of consequences of this account of Frege's. Since there is no King of
France, the phrase 'the King of France' would, for Frege, denote the
null class. Again take '$\Phi(\xi)$' to mean 'ξ is a son of Mr. So-and-so'. If
Mr. So-and-so has more than one son, then 'the only son of Mr.
So-and-so' = '$\iota\theta\Phi(\theta)$' would, for Frege, denote the set of sons of Mr.
So-and-so. Russell regards these consequences of Frege's theory as
"plainly artificial' and unsatisfactory. As he says, (1905) On Denot-
ing p. 47:

> "... Frege ... provides by definition some purely conventional denotation for the
> cases in which otherwise there would be none Thus 'The King of France', is to
> denote the null-class; 'the only son of Mr. So-and-so' (who has a fine family of ten),
> is to denote the class of all his sons; and so on. But this procedure, though it may

not lead to actual logical error, is plainly artificial, and does not give an exact analysis of the matter."

Frege's conventional denotations are indeed somewhat artificial, but then to talk at all of the King of France or of the only son of Mr. So-and-so (who has a fine family of ten), is rather curious. It could be argued that ordinary language does not give any clear sense to these strange locutions, so that any prescriptions in such cases would of necessity be artificial; and that Frege's prescriptions are consequently no more artificial than alternative suggestions. Russell of course wants to replace Frege's account of '$\imath\theta\Phi(\theta)$' by his own theory of descriptions — though this is not something which we will consider here.

Frege's basic law VI deals with $\imath\theta\Phi(\theta)$. It is

$$a = \imath\theta(a = \theta)$$

Frege has now sufficient logical apparatus at his disposal to translate his definition of number (as given in his (1884) Foundations of Arithmetic) into symbols, and to prove the corresponding theorems. This of course he proceeds to do. He deals first with many-one relations and one-one relations; next gives his definition of number in general, and of the particular numbers 0 and 1; and then goes on to define: "n follows in the series of natural numbers directly after m".

When discussing Frege's (1884) Foundations of Arithmetic, we listed three propositions which Frege states and whose proofs he there sketches informally. These propositions were (c.f. Ch. 7 p. 47):

(1) 'm is a successor of n' is a 1-1 relation.

(2) Every number except 0 is a successor of a number.

(3) Every number has a successor.

These propositions are proved formally in the *Grundgesetze*. (1) is Theorem 90; (2) is Theorem 107; while (3) is implied by Theorem 155.

In Vol. I of the *Grundgesetze* (1893), Frege takes his formal development of arithmetic as far as Theorem 348. In Vol II (1903), he continues as far as Theorem 484. He then breaks off his treatment of natural numbers in order to consider real numbers and how they should be defined. This section of the *Grundgesetze* is interesting because, among other things, it contains extended criticisms of formalist conceptions of mathematics. However we will not here consider it in detail.

It must have seemed to Frege, as he was finishing off Vol. II of his *Grundgesetze* in the summer of 1902, that he had successfully completed the project on which he had worked for nearly all his academic life. It was true that he had not, as he admitted in the Introduction to the *Grundgesetze* (see above Ch. 8 p. ...), carried the derivation of arithmetic as far as Dedekind had done in *Was sind und was sollen die Zahlen?* But Dedekind's work had been informal in character, whereas Frege had made the logic he used fully explicit and formalized all the proofs. Moreover it must have seemed to Frege that, to get as far as Dedekind, it was only a matter of continuing the deductions along the lines that he had already amply indicated. It was just at this moment of apparent success for Frege that disaster struck in the shape of Russell's paradox.

Russell discovered his paradox in 1901, and wrote to Frege about it in a letter dated 16 June 1902. Frege replied on 22 June 1902. Here are a few extracts from his letter. Frege (1902) Letter to Russell pp. 127-8:

> "Your discovery of the contradiction caused me the greatest surprise and, I would almost say, consternation, since it has shaken the basis on which I intended to build arithmetic. It seems, then, ... that my Rule V ... is false ... . I must reflect further on the matter. It is all the more serious since, with the loss of my Rule V, not only the foundations of my arithmetic, but also the sole possible foundations of arithmetic, seem to vanish. ... In any case your discovery is very remarkable and will perhaps result in a great advance in logic, unwelcome as it may seem at first glance."

Russell's paradox is most easily derived from the so-called axiom of comprehension.

$$(\exists y)(\forall x)(x \,\varepsilon\, y \leftrightarrow P(x) \,) - (*)$$

We have only to substitute $x \notin x$ (x is not a member of itself) for $P(x)$ to get:

$$(\exists y)(\forall x)(x \,\varepsilon\, y \leftrightarrow x \notin x)$$

Setting B (for Bertie) in place of y, we have

$$(\forall x)(\, x \,\varepsilon\, B \leftrightarrow x \notin x)$$

and so

$$B \,\varepsilon\, B \leftrightarrow B \notin B. \text{ A contradiction.}$$

To derive Russell's paradox in Frege's system, we have only to show how the axiom of comprehension follows from Frege's Basic Law V. In fact Frege proves a version of the axiom of comprehension as theorem 1 of the *Grundgesetze*. This is

$$f(a) \leftrightarrow a \,\varepsilon\, \grave{\theta}f(\theta) - (**)$$

Setting $a \notin a$ for $f(a)$ in (**), Russell's paradox follows as before. We have thus only to show how (**) follows from Frege's Basic Law V. The full formal derivation is given in Furth's translation of the *Grundgesetze* pp. 123-126. We will here sketch informally the basic ideas of the proof.

Suppose first that $a \; \varepsilon \; \hat{\theta}f(\theta)$. Then, by Frege's definition of the membership relation, it follows that there is a function g such that $g(a)$ and $\hat{\alpha}g(\alpha) = \hat{\theta}f(\theta)$. But Frege's Basic Law V states that

$$(\hat{\theta}f(\theta) = \hat{\alpha}g(\alpha) \;) \leftrightarrow (\forall\mathbf{a})(f(\mathbf{a}) \leftrightarrow g(\mathbf{a}) \;)$$

So we have

$$(\forall\mathbf{a})(f(\mathbf{a}) \leftrightarrow g(\mathbf{a}) \;)$$

and therefore in particular

$$f(a) \leftrightarrow g(a)$$

Since $g(a)$ holds, so does $f(a)$.

We have thus shown that $a \; \varepsilon \; \hat{\theta}f(\theta) \rightarrow f(a)$. Conversely, if $f(a)$ holds, then there is a function g such that $g(a)$ and $\hat{\alpha}g(\alpha) = \hat{\theta}f(\theta)$ – namely f itself. Therefore, by Frege's definition of the membership relation, $a \; \varepsilon \; \hat{\theta}f(\theta)$.

The news of Russell's paradox reached Frege too late for him to change the second volume of the *Grundgesetze*. However, he did add an appendix which states and discusses the contradiction. Here he writes (Furth translation. p. 127):

> "Hardly anything more unwelcome can befall a scientific writer than that one of the foundations of his edifice be shaken after the work is finished.
>
> I have been placed in this position by a letter of Mr. Bertrand Russell just as the printing of this [second] volume was nearing completion. It is a matter of my Basic Law (V). ...
>
> "*Solatium miseris, socios habuisse malorum.* (It is solace to the wretched, to have had companions in their misfortunes – D.G.) I too have this solace, if solace it is; for everyone who in his proofs has made use of extensions of concepts, classes, sets,[1] is in the same position. It is not just a matter of my particular method of laying the foundations, but of whether a logical foundation for arithmetic is possible at all."

Frege is quite correct to mention Dedekind in this connection since Dedekind, as we have seen (Ch. 8 p. 58), gives a more-or-less explicit formulation of the axiom of comprehension. Frege could also have mentioned Peano who presupposes the axiom of comprehension in his development of logic (c.f. Ch. 11 p. 77).

Frege goes on in the appendix to suggest a method by which Russell's contradiction might be resolved. His solution does not in fact work. Indeed new contradictions can be derived even when Frege's basic law V is emended in the way he suggests. This was apparently first shown by Leśniewski in 1938. A full account of Frege's proposed solution, and the new contradictions to which it leads is contained in the articles: W. V. Quine (1955) On Frege's Way Out, and P. T. Geach (1956) On Frege's Way Out.

---

[1] Herr R. Dedekind's 'systems' also come under this head. (Frege's footnote).

We ourselves will not, however analyse Frege's suggested solution, but rather stop at this point. The discovery of the contradictions marks a natural watershed in the study of the foundations of mathematics. The paradoxes raised new problems, and an analysis of the attempts to solve these problems lies beyond the limits of the present work.

Appendix I.
On Notation

In the text I have tried to use a fairly standard modern notation when writing formulas involving mathematical logic and set theory — except in quoting from older writers when I have sometimes retained the original notation. The aim of this appendix is to explain the modern notation employed.

1. Propositional Calculus

This deals with compound propositions formed out of simpler propositions by means of connectives. We shall use the following notation for connectives: $\neg$ = not, $\vee$ = or, $\wedge$ = and, $\leftrightarrow$ = if ... then, $\leftrightarrow$ = if and only if. We shall deal only with standard logic, and so it will be assumed that any proposition is either true or false, but not both (2-valuedness assumption). Granted this assumption, we can define the connectives in terms of 'true' and 'false' as follows (where p, q stand for arbitrary propositions)

$\neg$ p is true if and only if p is false

p $\vee$ q is false if and only if p is false and q is false

p $\wedge$ q is true if and only if p is true and q is true

p $\rightarrow$ q is false if and only if p is true and q is false

p $\leftrightarrow$ q is true if and only if p and q are either both true, or both false

2. Predicate Calculus

In the predicate calculus there are variables written e.g. $x, y, z, ..., x_1, x_2, ..., x_n, ...$ which are presumed to range over a domain of objects e.g. human beings, physical things, natural numbers, etc.

There are also a number of predicates:

1-place predicates $P(x)$ e.g. x is a man

2-place predicates (or relations) $R(x, y)$ e.g. x is to the left of y

...

n-place predicates $P(x_1, x_2, \ldots, x_n)$ e.g. $x_1 < x_2 < x_3 < \ldots < x_n$

There are in addition quantifiers:

'$(\forall x)P(x)$' means 'for all x, $P(x)$ holds'

'$(\exists x)P(x)$' means 'there is an x, for which $P(x)$ holds'

Since we are still making the assumption of 2-valuedness, either of these can be defined in terms of the other e.g.

$$(\exists x)P(x) =_{\text{def}} \neg(\forall x)\neg P(x)$$

or $\quad (\forall x)P(x) =_{\text{def}} \neg(\exists x)\neg P(x)$

If quantification is restricted to the variables x, y, z, ..., we speak of 1st-order predicate calculus. If quantification over the predicates is also allowed e.g. $(\forall P)(\forall x)(P(x) \to P(x)~)$, we speak of higher-order predicate calculus.

We shall use '$\imath x P(x)$' to mean 'the x such that $P(x)$'.

3. Set Theory

Let A be a *set*. 'a is a member of A', we shall write as '$a ~\varepsilon~ A$'. Let $P(x)$ be a predicate. Because of Russell's paradox (see Ch. 12 p. 91) we cannot automatically assume that the set of all x's such that $P(x)$ exists, but, if it does exist, we shall denote it by $\hat{x}P(x)$. We shall use the following notation for set-theoretic operations: $\subseteq$ = subset, U = union, $\cap$ = intersection. These may be defined as follows:

$A \subseteq B$ if and only if $(\forall x)(~(x ~\varepsilon~ A) \to (x ~\varepsilon~ B)~)$

$A ~U~ B =_{\text{def}} \hat{x}(~(x ~\varepsilon~ A) ~v~ (x ~\varepsilon~ B)~)$

$A \cap B =_{\text{def}} \hat{x}(~(x ~\varepsilon~ A) \wedge (x ~\varepsilon~ B)~)$

The set which has no members is called the *empty* or *null* set. We shall denote it by **ø**. It may be defined thus: **ø** $= \hat{x}(x \neq x)$.

Appendix II.
On the Principle of Mathematical, or Complete, Induction.

This principle states that if (α) a property P holds for 0, and (β) whenever the property holds for a natural number n, then it holds also for $n + 1$, then the property holds for all natural numbers. In symbols:

$(P(0) \wedge (\forall n)(P(n) \to P(n + 1))) \to (\forall n)P(n)$

If we take the natural numbers as beginning with 1 rather than 0, we merely substitute 1 for 0 in the above formulation.

The statement of the principle of mathematical, or complete, induction just given is in terms of properties or predicates; but we can also state the principle in terms of sets. Taking the natural numbers as beginning with 1, it becomes the following. Given any set S, if (α) 1 ε S, and (β) whenever a natural number $n \, \varepsilon$ S, then $n + 1 \, \varepsilon$ S, then all natural numbers are in S. In symbols:

$((1 \, \varepsilon \, S) \wedge (\forall n)(n \, \varepsilon \, S) \to (n + 1 \, \varepsilon \, S))) \to (\forall n)(n \, \varepsilon \, S)$

Suppose N $=$ the set of natural numbers
$= \{1, 2, 3, \dots, n, \dots\}$
Then we can write $(\forall n)(n \, \varepsilon \, S)$ as N $\subseteq$ S.

The principle of mathematical, or complete, induction is used constantly in mathematical developments of number theory. We shall illustrate thus by stating a very simple theorem, and proving it using mathematical induction.

Theorem $1 + 2 + 3 + \dots + n = \dfrac{n}{2}(n + 1)$

Proof (α) The theorem holds for $n = 1$, for, in this case $1 + 2 + 3 + \dots + n = 1$

and $\dfrac{n}{2}(n + 1) = \dfrac{1}{2} . 2 = 1$

So the two sides of the equation are equal.

(β) Suppose the theorem holds for n i.e. $1 + 2 + 3 + \dots + n$
$= \dfrac{n}{2}(n + 1)$

Then
$$1 + 2 + 3 \ldots + n + (n + 1) = \frac{n}{2}(n + 1) + (n + 1)$$
$$= \frac{(n + 1)(n + 2)}{2}$$

So if the theorem holds for n, it also holds for $n + 1$. Thus, by the principle of mathematical, or complete, induction, the theorem holds for all natural numbers n.

References

In the text works are referred to by the date of first publication and short title e.g. Frege (1879) *Begriffsschrift*. In what follows the full title and the exact edition from which quotations were taken are specified e.g. Frege, G., (1879) *Begriffsschrift, eine der arithmetischen nachgebildete Formelsprache des reinen Denkens*. English translation by T. W. Bynum as 'Conceptual Notation and related articles'. Oxford University Press. 1972.

Bell, J. L., and Machover, M., (1977) A Course in Mathematical Logic. North-Holland. 1977.

Boole, G., (1847) The Mathematical Analysis of Logic, being an Essay towards a Calculus of Deductive Reasoning. Basil Blackwell. 1965.

Brouwer, L. E. J., (1975) Collected Works. Vol. I. Philosophy and Foundations of Mathematics. ed. A. Heyting. North-Holland. 1975.

Chauvin, R., (1963) Animal Societies. English translation. Sphere books. 1971.

Currie, G., (1978) The Objectivism of Frege and Popper: an Historical and Critical Investigation. University of London. Ph. D. Thesis. (Unpublished).

Dedekind, R., (1872) Continuity and Irrational Numbers. English translation in 'Essays on the Theory of Numbers' Open Court. 1909.

— (1888) *Was sind und was sollen die Zahlen?* English translation in 'Essays on the Theory of Numbers' Open Court. 1909.

Fraenkel, A. A., and Bar-Hillel, Y., (1958) Foundations of Set Theory. North-Holland. 1958.

Frege, G., (1879) *Begriffsschrift, eine der arithmetischen nachgebildete Formelsprache des reinen Denkens.* English translation by T. W. Bynum as 'Conceptual Notation and related articles'. Oxford University Press. 1972.

— (1882) On the Aim of the *Begriffsschrift*. English translation in T. W. Bynum (ed.) Conceptual Notation and related articles. Oxford University Press. 1972. pp. 90-100.

— (1884) The Foundations of Arithmetic. A logico-mathematical enquiry into the concept of number. English translation by J. L. Austin. Basil Blackwell. 1968.

— (1892) On Sense and Reference, in P. Geach and M. Black (eds) Translations from the Philosophical Writings of Gottlob Frege. Blackwell. 1960. pp. 56-78.

— (1893) *Grundgesetze der Arithmetik, Begriffsschriftlich abgeleitet.* Vol. I. English translation of §§ 0-52 + Appendices by M. Furth as 'The Basic Laws of Arithmetic. Exposition of the System.' University of California. 1964.

— (1893 & 1903) *Grundgesetze der Arithmetik, Begriffsschriftlich abgeleitet.* Vol. I. (1893) and Vol. II. (1903). Reprinted by G. Olms. 1962.

— (1895) A Critical Elucidation of some Points in E. Schröder's *Vorlesungen über die*

Algebra der Logik. English translation in P. Geach and M. Black (eds). Translations from the Philosophical Writings of Gottlob Frege. Blackwell. 1960. pp. 86-106.

— (1896) *Über die Begriffsschrift des Herrn Peano und meine eigene,* in G. Frege. Kleine Schriften. Herausgegeben von Ignacio Angelelli. Georg Olms Verlagsbuchhandlung. Hildesheim. 1967. pp. 220-233.

— (1902) Letter to Russell. English translation in J. van Heijenoort (ed.) From Frege to Gödel. Harvard University Press. 1967. pp. 127-8.

— (1918) Thoughts. English translation by P. T. Geach and R. H. Stoothoff in P. T. Geach (ed.) Logical Investigations by F. Frege. Basil Blackwell. 1977. pp. 1-30.

Geach, P. T., (1956) On Frege's Way Out, in E. D. Klemke (ed.) Essays on Frege. University of Illinois Press. 1968. pp. 502-4.

Grenville, J. A. S., (1976) Europe Reshaped. 1848-1878. Fontana/Collins. 1976.

Kant, I., (1781) Critique of Pure Reason. English Translation by Norman Kemp Smith. Macmillan. 1958.

— (1783) Prolegomena to any Future Metaphysics that will be able to present itself as a Science. English translation by Peter G. Lucas. Manchester University Press. 1959.

Kline, M., (1972) Mathematical Thought from Ancient to Modern Times. Oxford University Press. 1972.

Lukasiewicz, J., (1934) On the History of the Logic of Propositions, in S. McCall (ed.) Polish Logic 1920-1939. Oxford. 1967. Ch. 4. pp. 66-87.

Mill, J. S., (1843) A System of Logic Ratiocinative and Inductive being a connected view of the principles of evidence and the methods of scientific investigation. 8th Edition. Longmans. 1936.

Peano, G., (1889) *Arithmetices principia nova methodo exposita.* English translation in H. C. Kennedy (ed.) Selected Works of Giuseppe Peano. Allen & Unwin. 1973. pp. 101-134.

— (1891) *Sul concetto di numero,* in U. Cassina (ed.) Opere Scelte. Rome. 1958. Vol. III. pp. 80-109.

— (1894) *Notations de Logique Mathematique. Introduction au Formulaire de Mathematiques,* in U. Cassina (ed.) Opere Scelte Rome. 1958. Vol. II. (66) pp. 123-175.

Popper, K. R., (1972) Objective Knowledge: an Evolutionary Approach. Oxford University Press. 1972.

Quine, W. V., (1955) On Frege's Way Out, in E. D. Klemke (ed.) Essays on Frege. University of Illinois Press. 1968. pp. 485-501.

Russell, B. A. W., (1902) Letter to Frege. English translation in J. van Heijenoort (ed.) From Frege to Gödel. Harvard University Press. 1967. pp. 124-5.

— (1905) On Denoting, in R. G. Marsh (ed.) Bertrand Russell. Logic and Knowledge. Essays 1901-1950. Allen & Unwin. pp. 41-56.

— (1919) Introduction to Mathematical Philosophy. Allen & Unwin. 1960.

— (1946) History of Western Philosophy and its Connection with Political and Social Circumstances from the Earliest Times to the Present Day. Allen & Unwin. 1948.

Schröder, E., (1880) Review of Frege's *Begriffsschrift.* English translation in T. W. Bynum (ed.) Conceptual Notation and Related Articles. Oxford University Press. 1972. pp. 218-232.

Venn, J., (1880) Review of Frege's *Begriffsschrift,* in T. W. Bynum (ed.) Conceptual Notation and Related Articles. Oxford University Press. 1972. pp. 234-235.

Wang, Hao, (1957) The Axiomatization of Arithmetic, Journal of Symbolic Logic, *22,* 1957, pp. 145-158.

Wittgenstein, L., (1953) Philosophical Investigations. Basil Blackwell. 1963.

Zermelo, E., (1904) Proof that every set can be well-ordered. English translation in J. van Heijenoort (ed.) From Frege to Gödel. Harvard University Press. 1967. pp. 139-141.

— (1908) Investigations in the foundations of set theory I. English translation in J. van Heijenoort (ed.) From Frege to Gödel. Harvard University Press. 1967. pp. 199-215.

Index

Errata

p. 12, line 12f.b.   'mathemtical' → 'mathematical'

p. 19, line 1. 'assume' → 'assumes'

p. 25, line 13. 'and the' → 'and then'

p. 34, line 18. 'bagueness' → 'vagueness'

p. 39, line 6. 'stayements' → 'statements'

p. 42, line 9. '(P(o) v ($\forall n$) …' → '(P(0) $\land$ ($\forall n$) …'

p. 54, line 25. 'domain-claculus' → 'domain-calculus'

p. 55, line 14. 'fictious' → 'fictitious'

p. 56, line 3 f.b. 'n, ,,,' → 'n, …'

p. 56, line 2 f.b. 'as set S' → 'a set S'

p. 80, line 17. 'Japanse' → 'Japanese'

p. 87, line 4. 'ξ-argument' → 'ζ-argument'

p. 87, line 5. 'ζ-argument' → 'ξ-argument'

p. 89, line 15 f.b. '$\Phi(\theta) = \Delta$' → '$\theta\Phi(\theta) = \Delta$'

p. 90, line 11. 'thought his' → 'though this'

p. 95, line 7. '$\neg(\exists x)\neg P(x)$' → '$\neg(\forall x)\neg P(x)$'

p. 95, line 8. '$\neg(\forall x)\neg P(x)$' → '$\neg(\exists x)\neg P(x)$'

Printed in the United States
by Baker & Taylor Publisher Services